Selected Titles in This Series

737 **Arjen Doelman, Robert A. Gardner, and Tasso J. Kaper,** A stability index analysis of 1-D patterns of the Gray-Scott model, 2002

736 **Wojciech Chachólski and Jérôme Scherer,** Homotopy theory of diagrams, 2002

735 **Martina Brück, Xi Du, Joonsang Park, and Chuu-Lian Terng,** The submanifold geometries associated to Grassmannian systems, 2002

734 **Michel Van den Bergh,** Blowing up of non-commutative smooth surfaces, 2001

733 **Milé Krajčevski,** Tilings of the plane, hyperbolic groups and small cancellation conditions, 2001

732 **Jan O. Kleppe, Juan C. Migliore, Rosa Miró-Roig, Uwe Nagel, and Chris Peterson,** Gorenstein liaison, complete intersection liaison invariants and unobstructedness, 2001

731 **Jesús Bastero, Mario Milman, and Francisco J. Ruiz,** On the connection between weighted norm inequalities, commutators and real interpolation, 2001

730 **Suhyoung Choi,** The decomposition and classification of radiant affine 3-manifolds, 2001

729 **Michael Grosser, Eva Farkas, Michael Kunzinger, and Roland Steinbauer,** On the foundations of nonlinear generalized functions I and II, 2001

728 **Laura Smithies,** Equivariant analytic localization of group representations, 2001

727 **Anthony D. Blaom,** A geometric setting for Hamiltonian perturbation theory, 2001

726 **Victor L. Shapiro,** Singular quasilinearity and higher eigenvalues, 2001

725 **Jean-Pierre Rosay and Edgar Lee Stout,** Strong boundary values, analytic functionals, and nonlinear Paley-Wiener theory, 2001

724 **Lisa Carbone,** Non-uniform lattices on uniform trees, 2001

723 **Deborah M. King and John B. Strantzen,** Maximum entropy of cycles of even period, 2001

722 **Hernán Cendra, Jerrold E. Marsden, and Tudor S. Ratiu,** Lagrangian reduction by stages, 2001

721 **Ingrid C. Bauer,** Surfaces with $K^2 = 7$ and $p_g = 4$, 2001

720 **Palle E. T. Jorgensen,** Ruelle operators: Functions which are harmonic with respect to a transfer operator, 2001

719 **Steve Hofmann and John L. Lewis,** The Dirichlet problem for parabolic operators with singular drift terms, 2001

718 **Bernhard Lani-Wayda,** Wandering solutions of delay equations with sine-like feedback, 2001

717 **Ron Brown,** Frobenius groups and classical maximal orders, 2001

716 **John H. Palmieri,** Stable homotopy over the Steenrod algebra, 2001

715 **W. N. Everitt and L. Markus,** Multi-interval linear ordinary boundary value problems and complex symplectic algebra, 2001

714 **Earl Berkson, Jean Bourgain, and Aleksander Pełczynski,** Canonical Sobolev projections of weak type $(1, 1)$, 2001

713 **Dorina Mitrea, Marius Mitrea, and Michael Taylor,** Layer potentials, the Hodge Laplacian, and global boundary problems in nonsmooth Riemannian manifolds, 2001

712 **Raúl E. Curto and Woo Young Lee,** Joint hyponormality of Toeplitz pairs, 2001

711 **V. G. Kac, C. Martinez, and E. Zelmanov,** Graded simple Jordan superalgebras of growth one, 2001

710 **Brian Marcus and Selim Tuncel,** Resolving Markov chains onto Bernoulli shifts via positive polynomials, 2001

709 **B. V. Rajarama Bhat,** Cocylces of CCR flows, 2001

708 **William M. Kantor and Ákos Seress,** Black box classical groups, 2001

707 **Henning Krause,** The spectrum of a module category, 2001

(Continued in the back of this publication)

A Stability Index Analysis of 1-D Patterns of the Gray-Scott Model

MEMOIRS
of the
American Mathematical Society

Number 737

A Stability Index Analysis of 1-D Patterns of the Gray-Scott Model

Arjen Doelman
Robert A. Gardner
Tasso J. Kaper

January 2002 • Volume 155 • Number 737 (third of 5 numbers) • ISSN 0065-9266

American Mathematical Society
Providence, Rhode Island

2000 *Mathematics Subject Classification.*
Primary 35K57, 35B35, 35B25; Secondary 92E20, 35B32, 34C37, 34E15.

Library of Congress Cataloging-in-Publication Data

Doelman, A.

A stability index analysis of 1-D patterns of the Gray-Scott model / Arjen Doelman, Robert A. Gardner, Tasso J. Kaper.

p. cm. — (Memoirs of the American Mathematical Society, ISSN 0065-9266 ; no. 737)

"Volume 155, number 737 (third of 5 numbers)."

Includes bibliographical references.

ISBN 0-8218-2739-1 (alk. paper)

1. Reaction-diffusion equations. 2. Perturbation (Mathematics) 3. Eigenvalues. I. Gardner, Robert A., 1950– II. Kaper, Tasso J., 1964– III. Title. IV. Series.

QA3.A57 no. 737
[QA377]
510 s—dc21
[515′.353] 2001045832

Memoirs of the American Mathematical Society

This journal is devoted entirely to research in pure and applied mathematics.

Subscription information. The 2002 subscription begins with volume 155 and consists of six mailings, each containing one or more numbers. Subscription prices for 2002 are $524 list, $419 institutional member. A late charge of 10% of the subscription price will be imposed on orders received from nonmembers after January 1 of the subscription year. Subscribers outside the United States and India must pay a postage surcharge of $31; subscribers in India must pay a postage surcharge of $43. Expedited delivery to destinations in North America $35; elsewhere $130. Each number may be ordered separately; *please specify number* when ordering an individual number. For prices and titles of recently released numbers, see the New Publications sections of the *Notices of the American Mathematical Society.*

Back number information. For back issues see the *AMS Catalog of Publications.*

Subscriptions and orders should be addressed to the American Mathematical Society, P. O. Box 845904, Boston, MA 02284-5904. *All orders must be accompanied by payment.* Other correspondence should be addressed to Box 6248, Providence, RI 02940-6248.

Memoirs of the American Mathematical Society is published bimonthly (each volume consisting usually of more than one number) by the American Mathematical Society at 201 Charles Street, Providence, RI 02904-2294. Periodicals postage paid at Providence, RI. Postmaster: Send address changes to Memoirs, American Mathematical Society, P. O. Box 6248, Providence, RI 02940-6248.

This publication is indexed in *Science Citation Index®*, *SciSearch®*, *Research Alert®*, *CompuMath Citation Index®*, *Current Contents®/Physical, Chemical & Earth Sciences.*

Printed in the United States of America.

♾ The paper used in this book is acid-free and falls within the guidelines established to ensure permanence and durability.

Visit the AMS home page at URL: `http://www.ams.org/`

10 9 8 7 6 5 4 3 2 1 07 06 05 04 03 02

Contents

Abstract

In this work, we study the stationary one-pulse solutions of the Gray-Scott model, which consists of a coupled pair of singularly perturbed reaction-diffusion equations. In a previous study, we used the method of matched asymptotic expansions to derive formal stability results for these solutions. The scaling regimes in which these solutions are stable, respectively unstable, were identified by reducing the fourth-order linearized system to a second-order nonlocal eigenvalue problem, referred to as the NLEP.

In this paper, we analyze the perturbed equations using the Evans function and an associated topological invariant, known as the stability index. Two novel features arise in the application of these methods to the Gray-Scott model, that also arise in other reaction-diffusion systems. First, the fast singular limits of these solutions are approximated by strongly unstable solutions of Fisher's equation. Hence, from the perspective of a fast-slow decomposition of the full eigenvalue problem, one expects that there will always be an $\mathbf{O}(1)$ unstable eigenvalue. Nevertheless, there exists a broad region in the parameter space in which these solutions are stable. We present an analysis of and explanation for this apparently paradoxical behavior, by showing how, in the fast-slow decompositions of both the Evans function and the stability index calculation, the NLEP possesses a singularity which cancels the contribution from the apparent Fisher eigenvalue. The NLEP properly measures the coupling of the slow field to the fast field, and the strength of this coupling plays a central role in the solution's stability.

Second, in order to determine the precise multiplicity of *critical* eigenvalues, *i.e.*, eigenvalues that are close to the origin in the singular limit, it is necessary to perform index calculations based on an analytic continuation of the Evans function to a two-sheeted Riemann surface. This is due to the presence of a branch point of the usual Evans function which lies in the essential spectrum of the pulse and which converges toward the origin in the singular limit. Branch points of the Evans function inside the essential spectrum occur in many other important applications, and this has frequently proven to be a critical obstacle to the implementation of these methods in such situations. In particular, the previous construction of a related topological invariant called the stability index, is not generally valid inside the essential spectrum. We present a new and general construction of this index which is valid on the entire domain of analyticity of the continuation of the Evans function.

Received by the editor August, 1998.

1991 *Mathematics Subject Classification.* Primary 35K57, 35B35, 35B25; Secondary 92E20, 35B32, 34C37, 34E15.

Key words and phrases. reaction-diffusion equations, singular perturbation theory, nonlocal eigenvalue problem, Evans function, stability index, zero-pole cancellation, Hopf bifurcation, activator-inhibitor systems, pattern formation, Gray-Scott model.

Preface

The possibility of performing index calculations for the analytic continuation of the Evans function to a Riemann surface was first noted in [**14**]. This procedure is concurrently being implemented by Kevin Zumbrun and the second author in connection with the stability analysis of the weak limit of nonclassical viscous shock profiles, where similar issues regarding the presence of branch points arise. The authors gratefully acknowledge Kevin Zumbrun's contributions in this regard.

T.K. was partally supported by National Science Foundation CAREER grant DMS-9624471 and by an Alfred P. Sloan Foundation Research Fellowship while this work was carried out.

Arjen Doelman, Rob Gardner and Tasso Kaper

CHAPTER 1

Introduction

1.1. The Gray-Scott Model: Background

The irreversible Gray-Scott equations,

$$\begin{aligned} U_t &= \Delta U - UV^2 + A(1-U) \\ V_t &= D_V \Delta V + UV^2 - BV \end{aligned}$$

arise as a model for a class of irreversible autocatalytic chemical reactions [**15, 16, 17**]. The variables $U = U(\mathbf{x}, t)$ and $V = V(\mathbf{x}, t)$ denote the concentrations of two chemical species $\mathcal{U}$ and $\mathcal{V}$, the parameter D_V is the normalized diffusivity of $\mathcal{V}$, A denotes the rate at which $\mathcal{U}$ is fed from the reservoir into the reactor, and B is the overall rate of decay of $\mathcal{V}$. The equations support many interesting spatio-temporal patterns, such as self-replicating pulses and spots in 1- and multi-D, respectively, [**24, 25, 26, 8**]. As in [**25, 26, 8, 6**] we focus in this paper on the one dimensional equation with A, B and D_V small. Following [**8, 6**] we introduce $\delta \ll 1$ by setting $D_V = \delta^2$. Parameters A and B are assumed to scale as $A = a\delta^2$ and $B = b\delta^\beta$ where $a, b = \mathbf{O}(1)$ with respect to δ and $0 \leq \beta < 1$:

$$\begin{aligned} U_t &= U_{xx} - UV^2 + a\delta^2(1-U) \\ V_t &= \delta^2 V_{xx} + UV^2 - b\delta^\beta V \end{aligned} \tag{1.1}$$

Apart from self-replicating pulses, this equation also exhibits stable homoclinic and periodic stationary solutions with pulses which appear as spikes in the singular limit as $\delta \to 0$. For the latter solutions, the support of the V component shrinks to a thin layer as the amplitude of V tends to infinity. The existence of such solutions was rigorously established in [**8**] when the parameter β lies in the range $0 \leq \beta < 1$, and nonexistence was also established whenever $\beta > 1$. The transitional regime when $\beta = 1$ is analyzed in [**6**]. The nonexistence of 1-pulses seems to coincide with the onset of the pulse-splitting phenomenon.

The scalings of the parameters A, B and D_V in (1.1) are inspired by the simulations reported in [**25, 26, 8**]. A detailed motivation and derivation is presented in [**8**]. Homoclinic solutions of the type studied in this paper exist in a much wider region of the (A, B, D_V)-parameter plane, see [**4, 5**]. Moreover, the stability of these orbits can be determined with the methods presented in [**6**] and this paper. Other regions of the parameter plane have been studied in [**18**] and [**22**].

In this paper, we present a rigorous stability analysis of the 1-pulse homoclinic solutions. This complements a previous asymptotic analysis of the linearized problem [**6**], in which the eigenvalue problem for the full system was shown to be governed by scalar, nonlocal eigenvalue problem (see (NLEP) below). In particular, we present a complete justification of the formal predictions of (NLEP) regarding the stability and instability of the 1-pulse solutions near each of the three singular limits identified in [**6**].

In order to carry out the analysis, two key issues had to be addressed which distinguish the present problem:

- In the layer in which the fast dynamics occur, the u-component of the solution is the constant $\bar{c} = 3b\sqrt{b/a}$ to leading order, while the v-component is approximated by the homoclinic stationary solution of the scalar Fisher equation,

$$0 = v_{\xi\xi} + \bar{c}v^2 - bv, \tag{1.2}$$

  where (u, v) are suitable rescalings of (U, V), see (1.5). It is well known that the homoclinic solution $S(\xi)$ of (1.2) is strongly unstable due to the presence of an eigenvalue $\lambda_1 > 0$ of the linearization (1.14) of (1.2) about $S(\xi)$. Nevertheless, both numerical and asymptotic analyses of the full system (1.1) predict that the exact stationary solutions of the full system in the fast scaling, see (1.6) below, will sometimes be stable for appropriate ranges of parameters. How can a solution constructed by matched asymptotics, in which the approximating solution of the inner (fast) layer has an $\mathbf{O}(1)$ *unstable* eigenvalue, paradoxically attain stability as a solution of the full system?
- An important issue in singular limit problems is to determine the multiplicity and location of all eigenvalues that are close to zero for $\delta << 1$. However, in the present case, the essential spectrum of the wave converges towards the translational eigenvalue at $\lambda = 0$ in the singular limit as $\delta \to 0$. This is a commonly encountered complication for singular limits of reaction-diffusion systems.

The analysis of the NLEP and the resolution of the first issue can be stated in terms of a certain analytic function of the eigenvalue parameter λ called the *Evans function* $D(\lambda, \delta)$, and an associated topological invariant called the *stability index*. The latter is an integer invariant associated to a certain vector bundle over S^2 which we denote by $\mathcal{E}(K, \delta)$, which counts the multiplicity of eigenvalues of the wave interior to some simple closed contour $K \subset \mathbb{C}$ which does not intersect the spectrum of the wave. The fibers of the bundle are determined by certain distinguished solutions of the linearized equations about the wave over the cylinder $(\xi, \lambda) \in \mathbb{R} \times K$.

The Evans function $D(\lambda, \delta)$ is (roughly) a Wronskian composed of certain other distinguished solutions whose roots coincide with the eigenvalues of the wave. A connection between $D(\lambda, \delta)$ and the topology of $\mathcal{E}(K, \delta)$ results from the observation that the Evans function is essentially the transition function for the bundle. The stability index, $c_1(\mathcal{E}(K, \delta))$ is the first Chern number of the bundle. For bundles over S^2 there is a simple characterization of c_1 in terms of the winding number of $D(K, \delta)$. The stability index therefore counts the multiplicity of eigenvalues of the wave interior to K. The relevant analysis is developed in detail in **[1]**.

In the present context, the bundle in question is a 2-plane bundle with fibers in $\mathbb{C}^4$. Near the singular limit, the bundle can frequently be decomposed into a Whitney sum of line bundles $\mathcal{E}(K, \delta) = \mathcal{E}_f(K, \delta) \oplus \mathcal{E}_s(K, \delta)$, where the summands are associated with certain distinguished "fast" and "slow" solutions of the linearized equations in $\mathcal{E}(K, \delta)$. It then turns out that the index of $\mathcal{E}(K, \delta)$ is the sum of the indices of each of the two summands. This approach was initiated in **[13]** in the setting of singularly perturbed predator-prey equations, and the present analysis is based upon the framework developed there. The strategy is to construct a homotopy from the subbundles of $\mathcal{E}(K,\delta)$ to certain reduced bundles, the indices of which are determined by the simpler reduced eigenvalue problems for the various singular limits of the linearized system.

In this paper, the fast summand $\mathcal{E}_f(K,\delta)$ is isomorphic to a bundle $\mathcal{E}_{fisher}(K)$ constructed from the linearization (1.14) about the homoclinic wave of the Fisher equation (1.2). This wave is well known to have a strongly unstable eigenvalue $\lambda_1 > 0$, and if K is a contour enclosing λ_1 in its interior, the fast subbundle must have index +1. Moreover, the analyticity of the Evans function *suggests* that the slow summand $\mathcal{E}_s(K,\delta)$ has nonnegative index. Thus when K is a contour in $\{\text{Re } \lambda > 0\}$ which encloses λ_1 in its interior, this observation, suggests that the index $c_1(\mathcal{E}(K,\delta))$ must be at least one, and hence that the perturbed pulses must always be unstable.

It turns out that this line of reasoning is incorrect because the transition function for the slow summand $\mathcal{E}_s(K,\delta)$ need not be analytic inside K. This is because one of the particular solutions used to generate the direct sum decomposition of the bundle is distinct from the corresponding one used to generate the Evans function. The latter solutions are analytic on and inside K, while the former solutions need not be. The reason this can occur is that the solution associated to $\mathcal{E}_s(K,\delta)$ is approximated by a solution of the NLEP (see Chapter 1, section 3), and the nonlocal character of the approximating reduced eigenvalue problem leads to the formation of a pole inside K for both the NLEP solution and also, for the perturbed solution of the exact equations. The slow summand therefore has a negative index which cancels the positive index of the fast summand. The constraint of analyticity of the Evans function for the full bundle implies that this pole must occur precisely at the point where the root in the transition function of fast summand occurs. It is this rather extraordinary cancellation that creates the possibility of a stable pulse in the Gray-Scott model. Similar pulse-stabilizing zero-pole cancellations occur in a broad class of reaction-diffusion systems (see ([**7**]).

The second difficulty alluded to above is further complicated because the Evans function of the wave has a branch point $\lambda_*(\delta) < 0$ on the boundary of the essential spectrum which converges to the origin as $\delta \to 0$. The domain of the Evans function as constructed in [**1**] is the slit plane $S_1 = \mathbb{C} \setminus \{\lambda \leq \lambda_*(\delta)\}$. The bundle approach as developed in [**1, 13**] requires that $K \subset S_1$. If the objective is to count eigenvalues close to the origin, the contour K must enclose the origin in its interior. Thus, K must intersect a point between $\lambda_*(\delta)$ and $\lambda = 0$. In such situations a key tracking lemma (Theorem 3.7, below) used in obtaining the Whitney sum decomposition of $\mathcal{E}(K,\delta)$ breaks down because K is asymptotically close to the translational eigenvalue at $\lambda = 0$.

It is therefore necessary to extend the analysis into the essential spectrum, and therefore, into the region in which $D(\lambda,\delta)$ becomes multivalued. The resulting object of study is the analytic continuation of the standard Evans function $D(\lambda,\delta)$ with domain S_1 to a 2-sheeted Riemann surface $\mathcal{R}$ with branch cut along the ray $\lambda < \lambda_*(\delta)$ along the negative real axis. However, the usual construction of the Evans function breaks down as soon as λ enters the essential spectrum. A similar problem also arises in the study of viscous undercompressive shock profiles [**14**]. The Gap Lemma in [**14**] provides a general procedure for analytically continuing $D(\tilde{\lambda},\delta)$ to a portion of the second sheet of $\mathcal{R}$ and hence, into the essential spectrum. A similar result was also recently proved in connection with the stability of travelling waves for a perturbed NLS equation, see ([**21**]).

It turns out that even though the Evans function can be analytically continued into the essential spectrum, the bundle construction in [**1**] is *incorrect* whenever the contour K intersects a certain domain in the essential spectrum with negative "spectral gap" (see Chapter 2, section 2 for definitions). This point is discussed in more detail in the remark following Corollary 4.15. In Chapter 2, section 3, we present a somewhat different bundle construction which retains its connection with the generalized eigenvalue problem for the analytic continuation $D(\tilde{\lambda}, \delta)$ of the Evans function into the essential spectrum, even when K intersects the domain with negative spectral gap. The bundle construction in [**1**] is isomorphic to the construction presented here when K is contained in a domain with positive spectral gap. Hence, the new bundle construction appears to be a more natural one than the one presented in [**1**]. The construction given in Chapter 2, section 3 is proved at the level of generality of the Gap Lemma [**14**], and it does not depend on any special features of the Gray-Scott model.

The bundle construction is then used to determine the multiplicity of eigenvalues of the perturbed wave in a neighborhood of $\lambda = 0$. We stress that this is a new method of analysis that should be applicable to many other singularly perturbed reaction-diffusion systems where the essential spectrum is close to the origin. In particular, we mention a singular reaction-diffusion system arising as a model for nonlinear wave guides studied by Rubin [**27**] where similar issues arise. Indeed, the equations studied in [**27**] were scaled in a nonphysical manner precisely in order to avoid the difficulty of the essential spectrum that we have dealt with in the Gray-Scott system. The techniques developed here should provide a means for studying the physical scaling of the equations in [**27**].

1.2. Asymptotic Analysis of the Wave

We present a brief summary of the geometric singular perturbation analysis used in [**8**] to construct the stationary 1-pulse solutions of

$$\begin{aligned} 0 &= U_{oxx} - U_o V_o^2 + a\delta^2(1 - U_o) \\ 0 &= \delta^2 V_{oxx} + U_o V_o^2 - b\delta^\beta V_o, \end{aligned} \tag{1.3}$$

where $0 \le \beta < 1$. The solutions $(U_o(x,\delta), V_o(x,\delta))$ in question consist of slow segments (outer layers) on the semi-infinite intervals $x < 0$ and $x > 0$, separated by a fast segment (or inner spike layer) on a vanishingly small interval about $x = 0$. On both slow segments, the component $V_o(x,\delta)$ is exponentially small in δ, and hence the slow solutions may be written as $(U_o(x,\delta), 0)$ to all orders. Also, since the δ^2 term is then the leading order term in the first equation of (1.3) in the slow regimes, it follows that the functions U_o are given to leading order by:

$$\bar{U}(x) = 1 - (1 - \bar{C})e^{-\delta\sqrt{a}|x|} \tag{1.4}$$

as $|x| \to \infty$. The, as yet unknown, constant $\bar{C}$ is determined by geometric singular perturbation theory, as briefly stated below.

The fast limit is resolved by stretching the independent variable and by using a rescaling of the dependent variables:

$$x = \delta^{(1-\beta/2)}\xi, \quad U(x) = \delta^{3\beta/2}u(\xi), \quad V(x) = \delta^{-\beta/2}v(\xi), \tag{1.5}$$

yielding the rescaled system of differential equations

$$(1.6)\qquad \begin{aligned} u_{o\xi\xi} &= \delta^{2(1-\beta)}[u_o v_o^2 - a\delta^{2-\beta/2} + a\delta^{2+\beta}u_o] \\ v_{o\xi\xi} &= -u_o v_o^2 + bv_o. \end{aligned}$$

See [8] for the derivation of this scaling, and we recall that $0 \leq \beta < 1$. In [8], it was shown that the asymptotic smallness of $u_{o_{\xi\xi}}$ implies that u_o is constant to high order in the fast regime: $u_o \equiv \bar{c}$, and that the v_o equation is to leading order that of a planar Hamiltonian system.

The explicit formula for the leading order term in the inner layer is given by

$$(1.7)\qquad (\bar{c}, S(\xi)) = (3b\sqrt{b/a}, \sqrt{a/(4b)}\text{sech}^2(\sqrt{b}\xi/2)),$$

where we recall that $S(\xi)$ is also the homoclinic solution of the Fisher equation (1.2). This value of $\bar{c}$ (and hence that of $\bar{C} = \delta^{\frac{3\beta}{2}}\bar{c}$) was determined by first identifying the basic geometric structures in the four-dimensional phase space of the system (1.3) and then by applying the Fenichel geometric singular perturbation theory, see [**10**]. The plane $\mathcal{M} \equiv \{(u, u_\xi, v, v_\xi)|v = 0, v_\xi = 0\}$ is a two-dimensional normally hyperbolic manifold that is invariant under the system (1.3); and, it is classified as a slow manifold because solutions on it evolve slowly. The slow manifold $\mathcal{M}$ possesses three-dimensional stable and unstable manifolds that intersect each other transversely in a two-dimensional intersection surface, as is shown via an appropriate Melnikov calculation. The desired stationary one-pulse solution $(U_o(\xi,\delta), V_o(\xi,\delta))$ lies on this surface. It is the unique solution on this surface that is both forward and backward asymptotic to the saddle-saddle fixed point at $(U = 1, V = 0)$ on $\mathcal{M}$, and its v component has a pulse centered precisely at $\xi = 0$ that is given to leading order by the homoclinic orbit of the planar Hamiltonian system (1.2) with $u_o = \bar{c}$. During the $\xi-$interval corresponding to this fast pulse, the derivative u_ξ undergoes a small jump of size $\mathbf{O}(\delta^{1-\beta})$ (that may be obtained by integrating the leading order term $u_o v_o^2$). To leading order, this jump must be of the same size as the distance on $\mathcal{M}$, as measured by the jump in U_x (calculated from the right and left limits as $x \to 0$ using (1.4) and then converted into the fast scaling), between the restricted stable and unstable manifolds of the saddle point $(u = \delta^{-3\beta/2}, u_\xi = 0)$. The requirement of equality between these two jump discontinuities in the first derivative of u yields precisely the condition that uniquely determines $\bar{c}$. Finally, explicit calculations in [8] verified that the basepoints of the fibers whose unions make up the two-dimensional intersection surface form curves on $\mathcal{M}$ that transversely intersect these restricted stable and unstable manifolds, and hence the slow and fast segments could be hooked up smoothly, according to the Fenichel theory ([**10**]). Moreover, the transverse intersection occurs at zero wave speed due to the presence of certain symmetries in the nonlinearities, whereas generally one would expect the perturbed wave to be nonstationary. Complete details of the existence result can be found in [8].

1.3. The Nonlocal Eigenvalue Problem (NLEP)

In order to determine the stability of the 1-pulse solutions $(U_o(x,\delta), V_o(x,\delta))$ of the perturbed system with $\delta > 0$ and small, we consider the linearization of (1.1) about the wave. These equations determine how small perturbations in the wave of the form $e^{\lambda t}(U(x), V(x))$ will evolve. It has been verified from our matched asymptotic analysis in [**6**] that the only regimes in which bounded eigenfunctions can occur are those for which λ is of order δ^β or smaller (see Appendix A, **[6]**). We therefore introduce a scaling of the eigenvalue parameter $b\delta^\beta\lambda$ directly into the equations, which yields the (slow) linearized system

$$
\begin{aligned}
U_{xx} &= [V_o(x,\delta)^2 + \delta^2 a + b\delta^\beta\lambda]U + 2U_o(x,\delta)V_o(x,\delta)V \\
\delta^2 V_{xx} &= -V_o(x,\delta)^2 U + [-2U_o(x,\delta)V_o(x,\delta) + b\delta^\beta(\lambda+1)]V.
\end{aligned}
\tag{1.8}
$$

An important difference between the asymptotics of this system and that of the system (1.3) for the nonlinear wave is the presence of the $b\delta^\beta\lambda$ term in the first equation, which is of lower order than the $\mathbf{O}(\delta^2)$ term in the first equation in (1.3). For completeness, we also record the fast scaling of (1.8) under the change of variables (1.6) in the form of an eigenvalue problem for a differential operator L:

$$
L\begin{pmatrix} u \\ v \end{pmatrix} = \begin{pmatrix} b\delta^2\lambda u \\ b\lambda v \end{pmatrix}, \tag{1.9}
$$

where

$$
L\begin{pmatrix} u \\ v \end{pmatrix} \equiv \begin{pmatrix} u'' - \delta^{2(1-\beta)}[(v_o(\xi,\delta)^2 + \delta^{2+\beta}a)u + 2u_o(\xi,\delta)v_o(\xi,\delta)v] \\ v'' + v_o(\xi,\delta)^2 u + [2u_o(\xi,\delta)v_o(\xi,\delta) - b]v \end{pmatrix}
$$

In [**6**], we used the technique of matched asymptotic expansions to show that decaying solutions of (1.8) are formally approximated by solutions of a scalar second-order equation with a nonlocal term. We dubbed this reduced equation the nonlocal eigenvalue problem (NLEP):

$$
v_{\xi\xi} + [2\bar{c}S(\xi) - b(1+\lambda)]v = -c(v;\lambda,\delta)S(\xi)^2, \tag{1.10}
$$

where

$$
c(v;\lambda,\delta) = \frac{-\delta^{1-2\beta}\bar{c}^3}{3b^{3/2}\delta^{1-2\beta} + \bar{c}^2\sqrt{b\lambda}} \int_{-\infty}^{+\infty} S(\xi)v(\xi,\lambda,c)\,d\xi. \tag{1.11}
$$

After carrying out this reduction, we analyzed the NLEP as follows to determine the formal stability of the one-pulse homoclinic orbits and of the spatially periodic states. Since the nonlocal term $c(v;\lambda,\delta)$ in the NLEP equation is a constant, a preliminary problem is to first solve the local equation

$$
v_{\xi\xi} + [2\bar{c}S(\xi) - b(1+\lambda)]v = -\tilde{c}S(\xi)^2, \tag{1.12}
$$

where $\tilde{c}$ is an arbitrary constant. If λ is *not* an eigenvalue of the operator on the left hand side of (1.12), then by Fredholm theory there is a unique solution $v(\xi,\lambda,\tilde{c})$ which decays exponentially as $|x| \to \infty$. Moreover, since $S(\xi)$ is explicitly known to be a sech^2 function, the equation can be solved by transforming it to the hypergeometric equation. The latter is then explicitly solved by using Mathematica to carry out a symbolic calculation to find the solution in terms of hypergeometric functions. For each $\tilde{c}$ there is a unique solution $v = v(\xi,\lambda,\tilde{c})$, and by linearity, it follows that $v(\xi,\lambda,\tilde{c}) = \tilde{c}v_*(\xi,\lambda)$, where $v_*(\xi,\lambda) = v(\xi,\lambda,1)$. The solution of (1.10) is thus reduced to solving the equation

$$
c(v(\xi,\lambda,\tilde{c}),\lambda,\delta) = \tilde{c},
$$

where c is the integral operator in (1.11). When $\tilde{c} \neq 0$ it can be normalized to unity, yielding the integro-differential system

$$
\begin{aligned}
v_*'' + [2\bar{c}S(\xi) - b(1+\lambda)]v_* &= -S(\xi)^2 \\
f(\lambda,\delta) \equiv 1 - c(v_*(\xi,\lambda);\lambda,\delta) &= 0,
\end{aligned}
\tag{1.13}
$$

where $v_*(\xi, \lambda)$ is a bounded solution of the first equation in (1.13). This problem is clearly the same as the NLEP above whenever the integral operator is nonzero. Note that the eigenvalue $\lambda = 0$ associated to the translational symmetry is still an eigenvalue of the NLEP (1.10), (1.11), since the corresponding eigenfunction $S'(\xi)$ is odd, so that by (1.11), $c = \tilde{c} = 0$. However, it is no longer an eigenvalue of the normalized NLEP (1.13).

For all practical purposes, (1.13) can be regarded as explicitly solvable. In particular, the explicit solution $v_*(\xi, \lambda)$ of the first equation in (1.13) was determined by Mathematica in terms of hypergeometric functions. Mathematica was again used to explicitly evaluate the integral in $c(v_*(\xi, \lambda); \lambda, \delta)$. This yields an explicit, albeit unwieldy function of the various parameters a, b, δ and λ. The roots λ of the resulting algebraic equation (1.13) can then be numerically determined as functions of the other parameters $(a, b,$ and $\delta)$ in the equation using a root finder. For any fixed $\mathbf{O}(1)$ value of a and any fixed but small positive value of δ, the various roots $\lambda_{iN}(b)$ of (1.13), which are the eigenvalues of the NLEP, can be determined as functions of b.

There are three distinct regimes that can occur for $\beta \in [0, 1)$ and all $\mathbf{O}(1)$ values of b, which we describe below. The separation into the three cases arises from the constraint that b be an $\mathbf{O}(1)$ parameter. More generally, we could introduce a second, independent small parameter $B = b\delta^\beta$ as in the original scaling (1.1). The assumption that $\beta < 1$ is equivalent to $\delta/B << 1$. In practice one must assume some relation between δ and B in order to distinguish the various possible singular limits. The introduction of $\beta \in [0, 1)$ and $b = \mathbf{O}(1)$ into the equations serves this purpose. The regimes described below are of course connected to each other by permitting b to be of higher order. However, no significantly new phenomena are generated in this manner, and it is clearer to focus on the three distinct cases which can occur by requiring b to be $\mathbf{O}(1)$.

(i) Strong instability: $0 \leq \beta < 1/2$. In this case the nonlocal term in (1.10) is of higher order, and the governing equation is just the linearized system about the homoclinic wave of the scalar Fisher equation,

$$v'' + [2\bar{c}S(\xi) - b(1 + \lambda)]v = 0. \tag{1.14}$$

This equation always has a positive eigenvalue $\lambda_1 = 5/4$ for all positive parameter values, and the NLEP has a positive $\mathbf{O}(1)$ eigenvalue $\lambda_{1N}(b)$ which is asymptotically close to λ_1. In addition, the asymptotic analysis together with the hypergeometric approach in [6] shows that there is another positive eigenvalue, $\lambda_{0N}(b)$ of the NLEP which is asymptotically close to the origin. Hence for $0 \leq \beta < 1/2$ the wave is always unstable.

(ii) The critical regime: $\beta = 1/2$. In this case, the nonlocal term is $\mathbf{O}(1)$ for small δ. In order to describe the results, δ is fixed at a positive yet sufficiently small value, a is again fixed at an $\mathbf{O}(1)$ value relative to δ, and the various eigenvalues of the perturbed wave are expressed as functions of the remaining $\mathbf{O}(1)$ parameter b. It can be checked from the expressions (1.7), (1.11) that the nonlocal term is of order $b^{-1/2}$ for large b and that the roots of (1.12) are $\mathbf{O}(1)$ close to those described above in the first regime with $\beta < 1/2$. However there now exist $\mathbf{O}(1)$ values $b_H < b_C$ where the results are different.

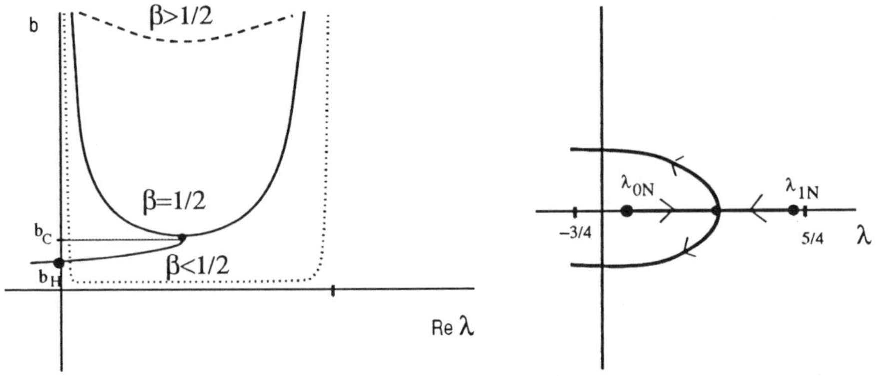

FIGURE 1.1. (a) Behavior of the real parts of eigenvalues of the NLEP as functions of b for various β. (b) Behavior of the eigenvalues of the NLEP as parametrized by b; arrows indicate the direction of decreasing b. Region B lies to the right of the dotted contour.

- For $b_C < b$ the wave is unstable with two positive eigenvalues $\lambda_{0N}(b) < \lambda_{1N}(b)$;
- For $b_H < b < b_C$ the wave is unstable with a pair of complex conjugate eigenvalues $\lambda_{0N}(b), \lambda_{1N}(b)$ with positive real part;
- For $b < b_H$, Re $\lambda_{iN}(b) < 0$ for $i = 0, 1$, and the translational eigenvalue at $\lambda = 0$ is the only eigenvalue in Re $\lambda \geq 0$.

(iii) Strong stability: $1/2 < \beta < 1$. In this case, the spectrum of the wave is precisely as in the third case $b < b_H$ in the transitional case $\beta = 1/2$, namely $\lambda = 0$ is simple and there are no unstable eigenvalues in $\{\text{Re } \lambda \geq 0\}$. The two eigenvalues $\lambda_{0N}(b)$ and $\lambda_{1N}(b)$ are complex conjugates with strictly negative real part for small δ.

It should be mentioned that there is a fourth regime which is distinct from the singular regimes considered in this paper that occurs when $\beta = 1$. The existence of pulses in this regime is studied in detail in [**6**], in which it is shown that there exists an $\mathbf{O}(1)$ value of $b = b_D$ for which the pulse exists for $b > b_D$ and for which no wave exists for $b < b_D$. Note that $\beta = 1$ is equivalent to scaling b by $\delta^{1-\beta}$ so that b is then of higher order for $\beta < 1$. The assumption here that b is an $\mathbf{O}(1)$ parameter for $\beta < 1$ implies that the disappearance value does not occur in the scalings considered in this paper. The stability of waves for $\beta = 1$ and for b larger than the "disappearance value" b_D has yet to be investigated.

The disappearance of the wave for very small b is closely linked to the emergence of the pulse-splitting phenomenon and the appearance of a stable periodic core; see [**6**]. The stability of the periodic waves constructed in [**8**] is therefore a topic of great interest. There are other interesting behaviors that can occur in this regime, such as the appearance of travelling waves with slowly varying wave speeds (see [**4, 5**]), a phenomenon which is also closely linked to pulse splitting.

In order to verify that the root finder indeed located all relevant solutions of (1.13), a numerical winding number calculation was performed for the analytic

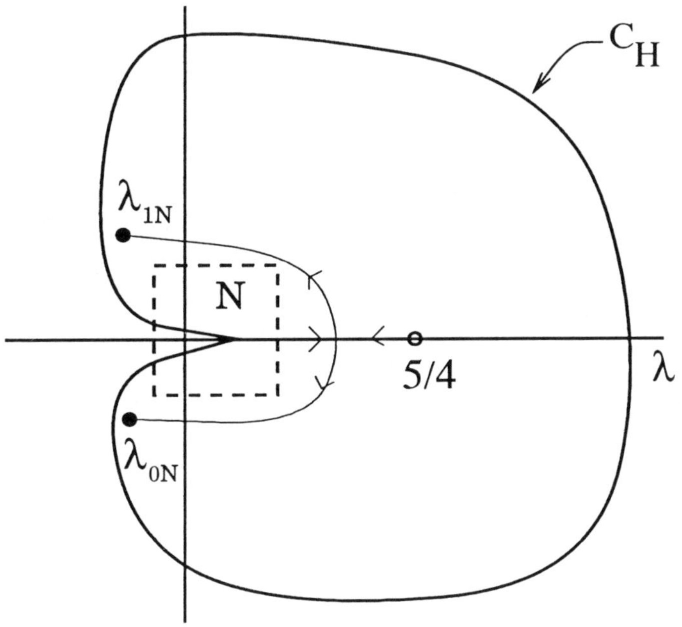

FIGURE 1.2. The contour C_H.

function

$$f(\lambda, \delta) = 1 - c(v_*(\xi, \lambda), \lambda, \delta) \tag{1.15}$$

along contours $C_H \subset \mathbb{C}$ which enclosed a large portion $|\lambda| \leq H$ of the unstable half plane in their interiors, where $H > 0$ is large but $\mathbf{O}(1)$ (see Figure 1.2).

For all parameter values, the two NLEP eigenvalues $\lambda_{0N}(b), \lambda_{1N}(b)$ located by the root finder remain in a uniformly bounded region of the spectral plane inside B, where B is defined as follows:

$$B = \{\lambda : d(\lambda, (-\infty, -3/4]) > 1/10\} \cap \{\lambda : \text{Re}\,\lambda > -11/10\}. \tag{1.16}$$

The excised interval along the negative real axis includes the essential spectrum of the Fisher wave $S(\xi)$, $\sigma_{ess}(S) = (-\infty, -1]$, together with an additional isolated eigenvalue of $S(\xi)$ at $\lambda = -3/4$. The region B has to be of this form since the two NLEP eigenvalues turn out to be near $-1.0 \pm 0.16i$ for $\beta > 1/2$.

In describing this computer-assisted portion of the program, we first remark that the asymptotic analysis in the Appendix of **[6]** determined that the only significant scaling for the eigenvalue parameter Λ in the scaling of the original equations (1.1) leading to the possibility of matched asymptotic expansions of bounded, decaying solutions of the linearized equations was the scaling $\Lambda = b\delta^\beta \lambda$ introduced directly into the linearized system (1.8), where λ remains $\mathbf{O}(1)$ as $\delta \to 0$ for each fixed $\beta \in [0, 1)$. For $0 \leq \beta \leq 1/2$ there is another scaling of Λ, namely $\Lambda = \mathbf{O}(\delta^{1-\beta}\lambda)$, for which matched bounded, asymptotic expansions of eigenfunctions can be constructed (see Appendix, **[6]**). However, since $1 - \beta \geq \beta$ for $\beta \leq 1/2$, such eigenvalues are asymptotically small in the significant scaling. There is exactly one such eigenvalue, however it is already accounted for in the solution of the NLEP, (i.e. $\lambda_{0N}(b)$). It is included inside the contour C_H and is one of the roots of $f(\lambda, \delta)$ found by the root finder. Hence for any other scaling of λ, there is no possibility of constructing a bounded eigenfunction near the singular limit.

For $|\lambda|$ large, we determined a simple explicit leading order representation of the function $f(\lambda, \delta)$. In particular, it is easily shown that the solution v_* of (1.13) is uniformly of order $|\lambda|^{-1}$ for large $|\lambda|$ in the right half plane. Thus the asymptotics of f determine an $\mathbf{O}(1)$ constant $H > 0$ such that $f(\lambda, \delta) \neq 0$ for $\lambda \in B \cap \{|\lambda| \leq H\}$.

As we shall see later, the region B intersects the continuous spectrum of the wave, which lies along a ray on the negative real axis whose endpoint is asymptotically close to the origin. This complicates the calculation because the square root function in the denominator of the coefficient of c becomes multivalued there. The winding number calculation could be performed on a portion of a suitable Riemann surface, however it is simpler to avoid this difficulty here by selecting contours C_H as depicted in Figure 1.2, in which C_H encloses the two NLEP eigenvalues $\lambda_{0N}(b)$ and $\lambda_{1N}(b)$ in its interior for all $b > 0$ and all $\beta \in [0, 1)$, and which *excludes* a small $\mathbf{O}(1)$ neighborhood of $\lambda = 0$ which is a subset of N, in the manner depicted in Figure 1.2.

Later in this paper, we prove an analytical result (see Lemma 5.3) in which it is shown that for $1/2 \leq \beta < 1$ there that there is a small but $\mathbf{O}(1)$ neighborhood N of the origin such that the perturbed wave has no eigenvalues interior to N other than the simple eigenvalue at $\lambda = 0$, for all sufficiently small $\delta > 0$. Hence for β in this range it is only necessary to determine the multiplicity of eigenvalues of the NLEP in the region interior to C_H, where C_H is as in Figure 1.2. The eigenvalue $\lambda_{0N}(b)$ is asymptotically small in the strong instability case, $0 \leq \beta < 1/2$. It turns up in the winding number calculation by choosing C_H appropriately.

One further complication is that the NLEP has a simple pole at $\lambda = 5/4$. This is because the Fisher wave has its principal eigenvalue at $\lambda = 5/4$, and so the differential equation in (1.13) is not solvable there. Hence, the correct winding number for $f(C_H, \delta)$ is $2 - 1 = 1$.

We calculated the winding number of $f(C_H, \delta)$ numerically with Mathematica for several typical parameter values, and in all cases the winding number was indeed unity. Hence the root finder successfully located all roots of the NLEP interior to C_H, while the asymptotic calculation ensures that a sufficiently large but fixed contour C_H will encircle all eigenvalues of the NLEP in the right half plane.

We have therefore given a numerical verification of the validity of the following proposition.

PROPOSITION 1.1. *(i) There exists an* $\mathbf{O}(1)$ *value* $H > 0$ *such that* $f(\lambda, \delta)$ *is uniformly bounded away from zero for* $\lambda \in B \cap \{|\lambda| \geq H\}$ *for all sufficiently small* $\delta > 0$ *and for* $\mathbf{O}(1)$ *values of* a, b.

(ii) For all $\mathbf{O}(1)$ *parameter values of* a, b *there are exactly two roots, namely* $\lambda_{0N}(b), \lambda_{1N}(b)$, *of* $f(\lambda, \delta)$ *for* $\lambda \in B$ *for all sufficiently small* $\delta > 0$.

A complete explanation of the reduction of (1.12) to the hypergeometric equation is given in **[6]**. The winding number calculation $W(f(\tilde{C}_H, \delta))$ was then performed with Mathematica, using the explicit formulae for $f(\lambda, \delta)$ appearing in **[6]** in terms of hypergeometric functions.

REMARK 1.2. In the following, it will be useful to distinguish the eigenvalues of the NLEP from those of the Fisher wave. We shall attach the subscript "N" whenever referring to eigenvalues of the NLEP. These eigenvalues are also functions of a, β and δ, however we shall suppress this dependence and only refer to them as functions of b. The positive eigenvalue $\lambda_1 = 5/4$ of the Fisher wave is actually independent of both a and b due to the way we have scaled λ.

1.4. The Main Results

The previous section outlines several formal stability and instability calculations based upon matched asymptotic expansions of the linearized equations. The objective here is to prove a theorem which validates the predictions of the matched asymptotics regarding stability and instability of the stationary 1-pulse solutions. The method of proof provides a truly new insight into the first key issue identified above, with attendant consequences for stability index techniques in general.

THEOREM 1.3. *Suppose that* $a = \mathbf{O}(1)$ *and fixed and that* b *is an* $\mathbf{O}(1)$ *parameter as* $\delta \to 0$. *Then there exists* $\delta_o > 0$ *such that for* $\delta < \delta_o$ *the multiplicity of eigenvalues of the operator* L *in (1.9) in the region* B *in (1.16) is exactly three. The translational eigenvalue at* $\lambda = 0$ *is always simple. The other two eigenvalues* $\lambda_0(b, \delta)$ *and* $\lambda_1(b, \delta)$ *are asymptotically close to the eigenvalues* $\lambda_{0N}(b)$ *and* $\lambda_{1N}(b)$ *of the NLEP for all* $\delta < \delta_0$. *In particular, the stability of the wave is as described in cases (i), (ii), and (iii) in the description of the NLEP in the previous section, i.e it is unstable for* $0 \leq \beta < 1/2$ *and stable for* $1/2 < \beta < 1$. *In case (ii) when* $\beta = 1/2$, *there is a transition to stability from instability as* b *is decreased through a small but* $\mathbf{O}(1)$ *neighborhood of* b_H.

REMARK 1.4. Since the linearized operator is sectorial and $\lambda = 0$ is simple, it follows from standard semigroup theory [**19**] that linearized stability implies nonlinear stability whenever the wave is linearly stable.

REMARK 1.5. The order of approximation of the perturbed eigenvalues $\lambda_i(b, \delta)$ by those of the NLEP $\lambda_{iN}(b)$ can be determined from the error analysis of Chapter 4 (see Theorem 4.11).

REMARK 1.6. As b crosses b_H, numerical simulations indicate that a subcritical Hopf bifurcation occurs, and the unstable stationary pulse continues to a stable stationary pulse together with a small amplitude unstable (temporally) periodic solution (see [**6**]). We have not attempted to justify the transversality criteria required for a rigorous proof that Hopf bifurcation occurs. It is likely that such information could be obtained with additional analysis of the parametric dependence of the explicit solution of the NLEP, the error analysis in case (ii) of Theorem 1.3, and the analyticity of the Evans function of the perturbed wave.

CHAPTER 2

The Evans Function and the Stability Index

2.1. The Linearized Equations

It will be convenient to express (1.8) as a first order, four-dimensional system

$$
\begin{aligned}
\dot{U} &= P \\
\dot{P} &= \Gamma_u U + 2U_o V_o V \\
\delta \dot{V} &= Q \\
\delta \dot{Q} &= -V_o^2 U + \Gamma_v V,
\end{aligned}
\tag{2.1}
$$

where "dot" is d/dx and

$$
\begin{aligned}
\Gamma_u(x,\lambda,\delta) &= b\delta^\beta \lambda + V_o(x,\delta)^2 + \delta^2 a \\
\Gamma_v(x,\lambda,\delta) &= b\delta^\beta \lambda - 2U_o(x,\delta)V_o(x,\delta) + b\delta^\beta.
\end{aligned}
$$

Setting $Y = (U,P,V,Q)^t$, (2.1) is expressed as

$$\dot{Y} = A(x,\lambda,\delta)Y,$$

where the A is the appropriate coefficient matrix. In the fast scaling (1.5), we obtain the equivalent system

$$
\begin{aligned}
u' &= \varepsilon\, p \\
p' &= \varepsilon\left[\gamma_u(\xi,\lambda,\delta)u + 2u_o(\xi,\delta)v_o(\xi,\delta)v\right] \\
v' &= q \\
q' &= -v_o(\xi,\delta)^2 u + \gamma_v(\xi,\lambda,\delta)v,
\end{aligned}
\tag{2.2}
$$

where $\varepsilon = \delta^{1-\beta}$, $' = d/d\xi$ and

$$
\begin{aligned}
\gamma_u(\xi,\lambda,\delta) &= v_o(\xi,\delta)^2 + \delta^{2\beta} b\lambda + \delta^{2+\beta} a \\
\gamma_v(\xi,\lambda,\delta) &= -2u_o(\xi,\delta)v_o(\xi,\delta) + b(1+\lambda).
\end{aligned}
\tag{2.3}
$$

We express (2.2) in vector form

$$y' = a(\xi,\lambda,\delta)y, \tag{2.4}$$

where $y = (u,p,v,q)^t$ and the coefficient matrix with entries evaluated at (ξ,λ,δ) is

$$
a(\xi,\lambda,\delta) = \begin{pmatrix} 0 & \varepsilon & 0 & 0 \\ \varepsilon\,\gamma_u & 0 & 2\varepsilon\, u_o v_o & 0 \\ 0 & 0 & 0 & 1 \\ -v_o^2 & 0 & \gamma_v & 0 \end{pmatrix}.
\tag{2.5}
$$

The asymptotic matrix for (2.2) is the limit of a as $|\xi| \to \infty$,

$$a_0(\lambda, \delta) = \begin{pmatrix} 0 & \varepsilon & 0 & 0 \\ \varepsilon \left(\delta^{2\beta} b\lambda + \delta^{2+\beta} a \right) & 0 & 0 & 0 \\ 0 & 0 & 0 & 1 \\ 0 & 0 & b(1+\lambda) & 0 \end{pmatrix}.$$

The structure of this matrix and of the associated o.d.e.'s $y' = a_0(\lambda, \delta)y$ plays an important role in studying the eigenvalue problem (2.4). We record the following (self-evident) facts for future reference.

LEMMA 2.1. *Let*

(2.6) $$\lambda_*(\delta) = -\delta^{2-\beta} a/b,$$

and let $S_1 = \mathbb{C} \setminus \{\lambda \in \mathbb{R} : \lambda \in (-\infty, \lambda_*(\delta)]\}$ *be the slit complex plane. Then for* $\lambda \in S_1$ *the eigenvalues of* $a_o(\lambda, \delta)$ *are given by*

$$\mu_1(\lambda) = \sqrt{b(1+\lambda)}; \qquad \mu_4(\lambda) = -\sqrt{b(1+\lambda)}$$
$$\mu_2(\lambda, \delta) = \delta\sqrt{b(\lambda - \lambda_*(\delta))}; \qquad \mu_3(\lambda, \delta) = -\delta\sqrt{b(\lambda - \lambda_*(\delta))},$$

where $\sqrt{\ }$ *denotes the positive branch of the square root with branch cut along the negative real axis. The associated eigenvectors are denoted by* $e_i(\lambda, \delta)$ *and are given by*

$$
\begin{aligned}
e_{1,4}(\lambda, \delta) &= (0, 0, 1, \mu_{1,4}(\lambda)^t \\
e_2(\lambda, \delta) &= (1, \delta^\beta \sqrt{b(\lambda - \lambda_*(\delta))}, 0, 0)^t \\
e_3(\lambda, \delta) &= (1, -\delta^\beta \sqrt{b(\lambda - \lambda_*(\delta))}, 0, 0)^t.
\end{aligned}
$$

It is important to note that the eigenvalues $\mu_{2,3}(\lambda, \delta)$ and the eigenvectors $e_{2,3}(\lambda, \delta)$ have a branch point at $\lambda = \lambda_*(\delta)$. The explicit formulae of Lemma 2.1 show that $\mu_{2,3}$ and $e_{2,3}$ are branches of analytic functions on the 2-sheeted Riemann surface $\mathcal{R}$ with branch cut along the half line $\lambda < \lambda_*(\delta)$.

2.2. The Evans Function

The Evans function $D(\lambda, \delta)$ for (2.2) is (informally) defined as a Wronskian of four distinguished solutions of (2.2), y_1 and y_2^* which decay to zero at $\xi = -\infty$, and y_3^* and y_4 which decay to zero at $+\infty$. The Evans function is then

(2.7) $$D(\lambda, \delta) = \det[y_1, y_2^*, y_3^*, y_4] \quad \text{at } (\xi, \lambda, \delta).$$

Note that since the trace of a in (2.4) is zero, $D(\lambda, \delta)$ is independent of ξ. We suppress its dependence on the other parameters in the equation for notational convenience.

In certain situations, it may not be possible to choose such a set of solutions analytically in λ, and it becomes necessary to define $D(\lambda, \delta)$ as a wedge of differential forms (see [1]). However, in the present situation, this is not the case, at least if Re $\lambda > \lambda_*(\delta)$. In particular, it follows from the above and the results in [1] that the following lemma holds.

LEMMA 2.2. *For $\lambda \in S_1$, there exist four solutions $y_{1,4}$, $y^*_{2,3}$ of (2.4) which satisfy*

$$\begin{aligned}\lim_{\xi\to-\infty}[y_1(\xi,\lambda,\delta)e^{-\mu_1\xi}] &= e_1(\lambda,\delta)\\ \lim_{\xi\to-\infty}[y_2^*(\xi,\lambda,\delta)e^{-\mu_2\xi}] &= e_2(\lambda,\delta)\\ \lim_{\xi\to+\infty}[y_3^*(\xi,\lambda,\delta)e^{-\mu_3\xi}] &= e_3(\lambda,\delta)\\ \lim_{\xi\to+\infty}[y_4(\xi,\lambda,\delta)e^{-\mu_4\xi}] &= e_4(\lambda,\delta).\end{aligned}$$

The solutions are analytic for $\lambda \in S_1$ and the convergence is uniform for λ in compact subsets of S_1. Furthermore, y_1 and y_4 are analytic and uniquely determined by the above conditions for all $\lambda \in \mathbb{C}\setminus(-\infty,-1]$.

PROOF. The lemma is obtained from [**1**] by noting that the "fast" eigenvalues $\mu_{1,4}(\lambda)$ have uniformly positive, negative real parts for such λ, so that Lemma 3.3 in [**1**] can be applied directly to obtain $y_{1,4}$ with the indicated behavior at $\pm\infty$. Since the real parts of the fast eigenvalues are strongly separated from the rest of the spectrum for Re $\lambda > -1$ for sufficiently small δ, the analyticity and uniqueness of $y_{1,4}$ for such λ follows as in [**1**]. Also, for $\lambda \in S_1$, the asymptotic matrix $a_o(\lambda,\delta)$ has 2-dimensional stable and unstable subspaces, and it again follows from [**1**] that there exist 2-dimensional subspaces $\Phi(\xi,\lambda,\delta)$ and $\Psi(\xi,\lambda,\delta)$ of solutions which are asymptotic to span$\{e_1,e_2\}$ as $\xi\to-\infty$ and to span$\{e_3,e_4\}$ as $\xi\to+\infty$. The two other solutions $y^*_{2,3}$ can be obtained from Φ and Ψ and the solutions $y_{1,4}$ by, for example, taking orthogonal complements of y_1 (resp. y_4) in Φ (resp. Ψ). □

REMARK 2.3. It is important to note that the asymptotic conditions for y_2^* and y_3^* in Lemma 2.2 are *not* sufficient to determine these solutions uniquely. Later, we shall make a different choice of the slow solutions $y_{2,3}$ which is distinct from the choice made in the Lemma. It turns out that this second choice is not globally analytic in λ. This point is of crucial importance in understanding the zero-pole cancellation that is implicit in the NLEP equation, and we will return to it later in Chapter 4 in our analysis of the slow solutions.

The following corollary characterizing the continuous spectrum of the linearized problem follows from Lemma 2.1 and some general functional analytic considerations (see e.g. [**19**], [**1**]).

COROLLARY 2.4. *The essential spectrum of the eigenvalue problem defined by the differential operator L in (1.9) lies along the negative real axis extending from $-\infty$ until $\lambda=\lambda_*(\delta)$ where $\lambda_*(\delta)$ is as in (2.6), and the spectrum of this system in the slit complex plane S_1 consists only of isolated eigenvalues of finite multiplicity.*

Thus the essential spectrum of L converges to the origin as $\delta\to0$. As described in Chapter 1, section 1, the solution to this problem is to analytically continue $D(\lambda,\delta)$ into the essential spectrum, and hence, to a portion of a 2-sheeted Riemann surface $\mathcal{R}$, where $\mathcal{R}$ is obtained by gluing two copies S_1, S_2 of the slit complex plane together along the ray $\{\lambda\le\lambda_*(\delta)\}$ in the usual way. Let $\tilde\lambda$ denote a generic point in $\mathcal{R}$. Note that the fast eigenvalues $\mu_{1,4}$ have branch points at $\lambda=-1$. These branch points play no role in the analysis, however they present a technical obstacle that needs to be avoided. To this end, let $B_1 = B\cap S_1$, where B is the domain in (1.16), and let $\mathcal{B}$ be the lifting of B_1 to the Riemann surface $\mathcal{R}$. It is immediate from Lemma 2.1 that the eigenvalues $\mu_i = \mu_i(\tilde{\lambda}, \delta)$ and eigenvectors $e_i(\tilde{\lambda}, \delta)$ all continue analytically to $\mathcal{B}$. The restriction to this portion of $\mathcal{R}$ is required in order to avoid a second branch point of the fast eigenvalues at $\lambda = -1$. Note that the analytic continuations of $\mu_{1,4}, e_{1,4}$ have the same value as the original ones on both sheets. This is not the case for $\mu_{2,3}, e_{2,3}$. We also note that the eigenspaces

$$\begin{aligned} \Phi(\tilde{\lambda}, \delta) &= \operatorname{span}\{e_1(\tilde{\lambda}), e_2(\tilde{\lambda}, \delta)\} \\ \Psi(\tilde{\lambda}, \delta) &= \operatorname{span}\{e_3(\tilde{\lambda}, \delta), e_4(\tilde{\lambda})\} \end{aligned} \tag{2.8}$$

of the asymptotic matrix $a_0(\tilde{\lambda}, \delta)$ continue analytically for $\tilde{\lambda} \in \mathcal{R}$, including the branch point.

Next, we show that the Evans function $D(\lambda, \delta)$ continues to an analytic function $D(\tilde{\lambda}, \delta)$ on some portion of $\mathcal{R}$ which includes the right half plane $\{\operatorname{Re} \lambda \geq 0\}$ on the first sheet of $\mathcal{R}$ together with a small but $\mathbf{O}(1)$ neighborhood $\tilde{N}$ of the branch point $\lambda_*(\delta)$ (see (2.6)) for all small $\delta > 0$. The difficulty is that the usual construction (see [**1**]) requires that the real parts of the eigenvalues associated to the eigenspace $\operatorname{span}\{e_1, e_2\}$ (resp. $\operatorname{span}\{e_3, e_4\}$) remain positive (resp. negative) for all λ in the domain of definition of $D(\lambda, \delta)$. This is not possible in $\tilde{N}$ since the real parts of μ_2 and μ_3 exchange signs when λ crosses the branch cut.

The Gap Lemma (see Theorem 2.3 in [**14**]) gives a different way of analytically continuing $D(\lambda, \delta)$ in this setting, provided that the extent of this crossing is not too large. One requirement of the Gap Lemma is that the subspaces $\Phi(\tilde{\lambda}, \delta)$ and $\Psi(\tilde{\lambda}, \delta)$ exist as analytic subspaces on the region of interest $\tilde{N}$ in the spectral plane. It follows from Lemma 2.1 that this is the case on the relevant portion of the Riemann surface.

A second requirement is that the *gap condition*

$$\Gamma(\tilde{\lambda}, \delta) > -\kappa. \tag{2.9}$$

holds in some region of $\mathcal{R}$. Here, $\kappa > 0$ is the exponential decay rate

$$|a(\xi, \lambda, \delta) - a_0(\lambda, \delta)| < Ke^{-\kappa|\xi|}$$

and $\Gamma(\tilde{\lambda}, \delta)$ is the *spectral gap* of $\Phi(\tilde{\lambda}, \delta)$ and $\Psi(\tilde{\lambda}, \delta)$. The spectral gap is

$$\Gamma(\tilde{\lambda}, \delta) = \operatorname{Re}(\mu_2(\tilde{\lambda}, \delta) - \mu_3(\tilde{\lambda}, \delta)), \tag{2.10}$$

The standard construction in [**1**] applies only to the region in the spectral plane for which the asymptotic systems maintain a *positive* spectral gap. This is called *the domain of consistent splitting*. Here, this region is the first sheet of $\mathcal{R}$ outside of the branch cut. The result in [**14**] gives another way to analytically continue $D(\lambda, \delta)$ into regions of the (generalized) spectral plane $\mathcal{R}$ for which (2.9) is satisfied. Since $\mu_{2,3}$ are both $\mathbf{O}(\delta)$, it suffices to show that the decay rate $\kappa > 0$ can be chosen independently of δ.

This would appear to be impossible, since the wave $(u_o(\xi, \delta), v_o(\xi, \delta))$ decays to its limits at the slow order δ exponential rate. However, it follows from the geometric construction in [8] that the v_o component of the wave decays at the uniformly *fast* exponential rate $e^{-\sqrt{b}|\xi|}$, since v_o is exponentially small along the slow manifolds. A curious structural feature of the Gray-Scott model is that the slow component u_o of the wave only enters into the linearized matrix $a(\xi, \delta, \lambda)$ as a factor of the fast component v_o (see (2.2)), and so we obtain a uniformly

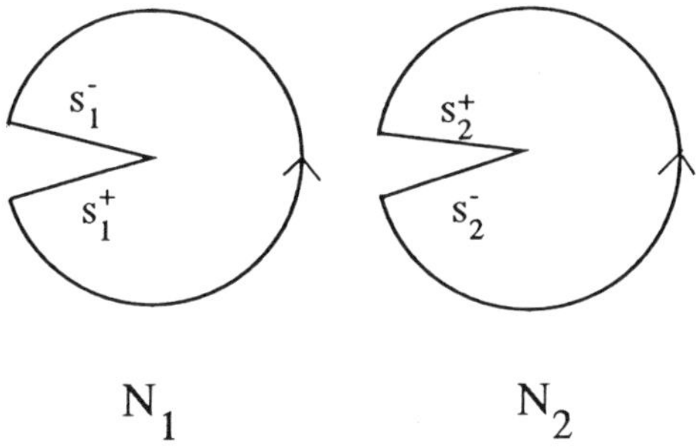

FIGURE 2.1. The two sheets of $\tilde{N}$

fast exponential decay rate of $|a - a_0| < Ke^{-\sqrt{b}|\xi|}$ for all $\tilde{\lambda}$. Hence $\kappa = \sqrt{b}$ and $\Gamma(\tilde{\lambda}, \delta) = \mathbf{O}(\delta)$, so that the gap condition (2.9) holds uniformly for $\tilde{\lambda} \in \tilde{N}$ and for all sufficiently small $\delta > 0$, where $\tilde{N}$ is the small but $\mathbf{O}(1)$ neighborhood of the branch point $\lambda_*(\delta)$ introduced above. The Gap Lemma therefore yields an analytic continuation $D(\tilde{\lambda}, \delta)$ of $D(\lambda, \delta)$ to the region

$$\mathcal{R}_+ = \tilde{N} \cup [B \cap S_1] \tag{2.11}$$

where B is the domain in (1.15). Since $\lambda_*(\delta)$ is asymptotically small, $\tilde{N}$ includes the origin $\lambda = 0$ on the first sheet of $\mathcal{R}$ for all sufficiently small δ.

We next characterize the neighborhood $\tilde{N}$ more explicitly. Let N be a small but $\mathbf{O}(1)$ neighborhood of $\lambda = 0$ in $\mathbb{C}$ and let $N_i = N \cap S_i$ for $i = 1, 2$. Then $\tilde{N}$ can be taken to be the neighborhood of $\lambda_*(\delta)$ in $\mathcal{R}$ obtained by gluing N_1 and N_2 together along the branch cut by attaching side s_1^+ (resp. side s_1^-) of N_1 to side s_2^+ (resp. s_2^-) in N_2, as depicted in Figure 2.1. Here, the superscript $+/-$ refers to the branch of the square root function which has the sign of $+/-$ along the positive real axis.

More precisely, a 2-cell (in real dimensions) is attached to each of the edges s_1^+, s_2^+ and s_1^-, s_2^-; since the surface lives in a (real) 4-dimensional space, the attached 2-cells can be chosen to lie in different hyperplanes, so that they intersect only at the branch point. The result is homeomorphic to a disk. Note that the branch point is in the interior of the disk.

The Gap Lemma then ensures the existence of analytic subspaces of solutions $\Phi(\xi, \tilde{\lambda}, \delta)$ and $\Psi(\xi, \tilde{\lambda}, \delta)$ of (2.2) which converge to $\Phi(\tilde{\lambda}, \delta)$ and $\Psi(\tilde{\lambda}, \delta)$ at $\xi = -\infty, +\infty$, respectively. There are in fact many ways to smoothly extend these spaces into the regions with negative spectral gap. The Gap Lemma selects a particular continuation which provides a *fast* exponential decay rate of the solution space to its geometric limit, and the continuation is analytic in $\tilde{\lambda}$. The fast exponential decay rate provides crucial information used later in Chapter 4 in the estimation of the slow solution y_2 solution inside the essential spectrum. We therefore give a more detailed description of the nature of this convergence theorem.

To this end, it is convenient to introduce a pair of 2-forms $\eta(\xi, \tilde{\lambda}, \delta)$ and $\zeta(\xi, \tilde{\lambda}, \delta)$ in $\Lambda^2\mathbb{C}^4$ whose spans are associated to the subspaces $\Phi(\xi, \tilde{\lambda}, \delta)$ and $\Psi(\xi, \tilde{\lambda}, \delta)$, respectively. On the first sheet of $\mathcal{R}$ these forms are simply $\eta = y_1 \wedge y_2^*$ and $\zeta = y_3^* \wedge y_4$, where the $y_{1,4}$ and $y_{2,3}^*$ are the solutions used in the definition of the Evans function. We can similarly associate the eigenspaces spaces $\Phi(\tilde{\lambda}, \delta)$ and $\Psi(\tilde{\lambda}, \delta)$ of $a_0(\tilde{\lambda}, \delta)$ to the span of 2-forms $\eta(\tilde{\lambda}, \delta)$ and $\zeta(\tilde{\lambda}, \delta)$. For definiteness, we take

$$\eta(\tilde{\lambda}, \delta) = e_1(\tilde{\lambda}) \wedge e_2(\tilde{\lambda}, \delta) \quad \zeta(\tilde{\lambda}, \delta) = e_3(\tilde{\lambda}, \delta) \wedge e_4(\tilde{\lambda}).$$

The Gap Lemma asserts that there is a *uniquely* determined pair of 2-forms $\eta(\xi, \tilde{\lambda}, \delta)$ and $\zeta(\xi, \tilde{\lambda}, \delta)$ of the associated linear system induced by (2.4) on $\Lambda^2\mathbb{C}^4$ which satisfy the exponential decay estimates

$$\begin{aligned} \eta(\xi, \tilde{\lambda}, \delta) &= e^{(\mu_1+\mu_2)\xi}(\eta(\tilde{\lambda}, \delta) + \mathbf{O}(e^{\tilde{\kappa}\xi})), \quad (\xi \to -\infty) \\ \zeta(\xi, \tilde{\lambda}, \delta) &= e^{(\mu_3+\mu_4)\xi}(\zeta(\tilde{\lambda}, \delta) + \mathbf{O}(e^{-\tilde{\kappa}\xi})), \quad (\xi \to +\infty), \end{aligned} \tag{2.12}$$

for any $\tilde{\kappa}$ satisfying $-\Gamma(\tilde{\lambda}, \delta) < \tilde{\kappa} < \kappa$.

We also observe that the solutions $y_1(\xi, \lambda, \delta)$ and $y_4(\xi, \lambda, \delta)$ are analytic in the neighborhood N_1 of the origin of the standard spectral plane, and so, they trivially continue analytically to all of $\tilde{N}$. Solutions $y^*_{2,3}(\xi, \tilde{\lambda}, \delta)$ can then be selected so that $y_1 \wedge y^*_2$ (resp. $y^*_3 \wedge y_4$) spans η (resp. ζ) and which are analytic in $\tilde{\lambda} \in \tilde{N}$. For example, we can simply take orthogonal complements of each y_1 (resp. y_4) in Φ (resp. Ψ) as in Lemma 2.2.

We summarize the above results in the following lemma

LEMMA 2.5. *There exists $\delta_o > 0$ such that for $\delta < \delta_o$ and $\tilde{\lambda} \in \mathcal{R}_+$ there exist 2-dimensional subspaces*

$$\begin{aligned} \Phi(\xi, \tilde{\lambda}, \delta) &\to \Phi(\tilde{\lambda}, \delta) \qquad (\xi \to -\infty) \\ \Psi(\xi, \tilde{\lambda}, \delta) &\to \Psi(\tilde{\lambda}, \delta) \qquad (\xi \to +\infty). \end{aligned}$$

The convergence is at a fast exponential rate (2.12) for any $\tilde{\kappa}$ satisfying

$$-\Gamma(\tilde{\lambda}, \delta) < \tilde{\kappa} < \sqrt{b}$$

for all $\tilde{\lambda} \in \tilde{N}$, where η and ζ are 2-forms representing $\Phi(\xi, \tilde{\lambda}, \delta)$ and $\Psi(\xi, \tilde{\lambda}, \delta)$, respectively. The fast decay rate (2.12) uniquely determines $\eta(\xi, \tilde{\lambda}, \delta)$ and $\zeta(\xi, \tilde{\lambda}, \delta)$, and these forms are analytic in $\mathcal{R}_+$ (see (2.11)).

The solutions $y_{1,4}$ and $y^*_{2,3}$ of Lemma 2.2 used in the definition of the Evans function have analytic continuations to $\tilde{N}$, and

$$D(\tilde{\lambda}, \delta) = \eta(\xi, \tilde{\lambda}, \delta) \wedge \zeta(\xi, \tilde{\lambda}, \delta) \tag{2.13}$$

provides an analytic continuation $D(\tilde{\lambda}, \delta)$ of $D(\lambda, \delta)$ to $\mathcal{R}_+$.

REMARK 2.6. The analytic continuation $D(\tilde{\lambda}, \delta)$ which extends $D(\lambda, \delta)$ to $\tilde{N}$ can be characterized explicitly as follows. On the first sheet of $\mathcal{R}$, where the standard branch of the square root is taken in the definition of $\mu_{2,3}(\lambda, \delta)$, $D(\tilde{\lambda}, \delta)$ is the standard Evans function $D(\lambda, \delta)$. This function measures the intersection of the subspace of solutions $\text{span}\{y_1, y^*_2\}$ of solutions of (2.4) which decay to zero at $\xi = -\infty$ with the subspace of solutions $\text{span}\{y^*_3, y_4\}$ of solutions which decay to zero at $+\infty$. Hence the roots of D on this sheet coincide precisely with the eigenvalues of the linearized operator L. On the second sheet of $\mathcal{R}$, $D(\tilde{\lambda}, \delta)$ still measures the intersection of solutions in the analytic continuation of these solution spaces, however the roots of $D(\tilde{\lambda}, \delta)$ are not necessarily eigenvalues of L, since there has been a switch in the signs of the real parts of two of the eigenvalues of a_0, $\text{Re}\,\tilde{\mu}_2 < 0 < \text{Re}\,\tilde{\mu}_3$, for $\tilde{\lambda}$ on the second sheet. Hence the solutions y^*_2 and y^*_3 grow at $-\infty$ and $+\infty$, respectively.

COROLLARY 2.7. *Let $\tilde{K} = \partial\tilde{N}$, and suppose that $D(\tilde{\lambda}, \delta) \neq 0$ for $\tilde{\lambda} \in \tilde{N}$. Let $n = W(D(\tilde{K}, \delta)) \geq 1$ be the winding number of $D(\tilde{K}, \delta)$ with respect to the origin. Then n is an upper bound for the multiplicity of eigenvalues of the pulse on the first sheet N_1 of $\tilde{N}$.*

The proof of the corollary follows immediately from the analyticity of $D(\tilde{\lambda}, \delta)$ on $\tilde{N}$ together with the above remark, the existence of the translational eigenvalue at $\lambda = 0$, and some standard complex analysis (see e.g. Theorem 2.10 below).

2.3. The Stability Index

The bundle construction in [**1**] is only valid, in regions of $\mathbb{C}$ in which the asymptotic systems have positive spectral gap. This construction exploits the fact that $\Phi(\xi, \lambda, \delta) \to \Phi(\lambda, \delta)$ whenever the spectral gap along K is positive and K is disjoint from the isolated eigenvalues of the wave. In situations where it is necessary to choose the contour K so that it intersects the region with negative spectral gap, as is the case in this paper, the construction in [**1**] is incorrect. The difficulty is that the geometric limit of $\Phi(\xi, \tilde{\lambda}, \delta)$ is discontinuous across the branch cut of $\mathcal{R}$ which separates the regions with positive and negative spectral gap; see the remark following Corollary 4.15 for further discussion.

Certain aspects of the new construction given below are still very similar to the approach in [**1**]. The principal ingredient is the subspace $\Phi(\xi, \lambda, \delta)$ of solutions of the linearized equations about the wave, as defined in the previous section, and the span of the 2-form $\eta(\xi, \lambda, \delta)$ representing this subspace and satisfying the condition (2.12). The span of this form determines a line bundle with fibers in $\Lambda^2\mathbb{C}^4$ and base space consisting of the cylinder $(\xi, \lambda) \in \mathbb{R} \times K$, where $K \subset S_1$ is a simple closed contour. The base space is then compactified through a change of variables $\xi = \xi(\tau)$ where $-1 < \tau < 1$; see (3.18) for an explicit formula. This reparametrization forms a bundle, $\text{span}\{\eta(\tau, \tilde{\lambda}, \delta)\}$, over the finite cylinder $B_* = (-1, 1) \times K$. A bundle over S^2 is constructed by gluing "caps" $B_\pm = \{\pm 1\} \times [K \cup K^{int}]$ on the ends of the cylinder B_* by defining the fibers over each of these caps to be span $\eta(\lambda, \delta)$. The continuity of the fibers at $\tau = -1$ follows from the fact that (2.12) is satisfied by $\eta(\xi, \tilde{\lambda}, \delta)$ at $\xi = -\infty$, while continuity at $\tau = +1$ results from the additional assumption K is disjoint from the spectrum of the wave (see [**1**]).

An alternative construction consists of truncating $\tau \leq \tau_+$ where $\tau_+ < 1$ but sufficiently close to one. The advantage of this approach is that it is not necessary to calculate the limit of the span of $\eta(\tau, \tilde{\lambda}, \delta)$ as $\tau \to +1$. However, it is no longer clear what subspace to glue on over the right cap at $\tau = \tau_+$. To complete the construction we employ a standard bundle procedure called the *clutch operation*, which we describe below.

We shall have occasion to construct bundles other than $\mathcal{E}(K, \delta)$, namely the fast and slow summands. Hence we present the new construction in a slightly more general context, which should also be useful in future applications of the theory.

Let $K = \tilde{K} \subset \mathcal{R}_+$ be a simple closed contour on a portion of $\mathcal{R}$ for which the gap condition (2.9) is satisfied. The base space $\mathcal{B}$ is now the union of $B_\pm$ and B_*, where $B_- = \{-1\} \times (\tilde{K} \cup \tilde{K}^{int})$, $B_* = (-1, \tau_+) \times \tilde{K}$, and $B_+ = \{\tau_+\} \times (\tilde{K} \cup \tilde{K}^{int})$. Suppose next that for $(\tau, \tilde{\lambda}) \in B_- \cup B_*$, we are given a pure k-form $\alpha(\tau, \tilde{\lambda}) \in \Lambda^k\mathbb{C}^N$ which represents a k-dimensional subspace $A(\tau, \tilde{\lambda})$ of $\mathbb{C}^N$. We assume that $\alpha(\tau, \tilde{\lambda})$

depends continuously on τ and analytically on $\tilde{\lambda}$. Furthermore, suppose that $\beta(\tilde{\lambda})$ is a given analytic k-form defined over the disk B_+.

A line bundle $\mathcal{E}(\tilde{K})$, with fibers in $\Lambda^k\mathbb{C}^N$ and base space $\mathcal{B} = B_- \cup B_* \cup B_+$ can then be constructed by the "clutching" operation (see **[2]**). To this end, consider the (trivial) line bundles $\mathcal{E}_-$ over the set $B_- \cup B_*$, and $\mathcal{E}_+$ over B_+ defined by taking the fibers over $B_- \cup B_*$ to be the span of $\alpha(\tau, \tilde{\lambda})$, and over B_+ to be the span of $\beta(\tilde{\lambda})$. The *clutch*, $\mathcal{E}(\tilde{K}) = \mathcal{E}_- \cup_F \mathcal{E}_+$ is obtained by forming the topological quotient space obtained by identifying points in the fibers of the restriction of $\mathcal{E}_\pm$ over the equator, $\{\tau_+\} \times \tilde{K}$ with a linear identification map $F(\tilde{\lambda})$. In order to define F, let $G_+(\tilde{\lambda})$ be a subspace of $\Lambda^k\mathbb{C}^N$ which is complementary to the space spanned by $\beta(\tilde{\lambda})$ and which depends analytically on $\tilde{\lambda}$. Under the above hypotheses, there is a well defined decomposition

$$\alpha(\tau_+, \tilde{\lambda}) = g(\tilde{\lambda})\beta(\tilde{\lambda}) + \gamma(\tilde{\lambda}), \tag{2.14}$$

where $g(\tilde{\lambda})$ is a scalar coefficient, and $\gamma(\tilde{\lambda}) \in G_+(\tilde{\lambda})$. Since the projection operation is analytic, both g and γ are then analytic functions of $\tilde{\lambda}$, for $\tilde{\lambda} \in \tilde{K}$.

Suppose next that $g(\tilde{\lambda}) \neq 0$ for $\tilde{\lambda} \in \tilde{K}$. The clutching map can then be defined by $F(\tilde{\lambda})\alpha(\tau_+, \tilde{\lambda}) = g(\tilde{\lambda})\beta(\tilde{\lambda})$ and extending linearly to get an isomorphism between the two subspaces. The resulting topological quotient space $\mathcal{E}(\tilde{K}) = \mathcal{E}_- \cup_F \mathcal{E}_+$ then forms a line bundle over the 2-sphere $\mathcal{B}$ with fibers in $\Lambda^k\mathbb{C}^N$, and the isomorphism class of $\mathcal{E}(\tilde{K})$ depends only on the homotopy type of the clutching function, $[g] \in \pi_1(S^1)$; see Lemma 1.4.7 in **[2]**.

The bundle $\mathcal{E}(\tilde{K})$ determines an isomorphic k-plane bundle through the Plücker embedding that associates a k-dimensional subspace of $\mathbb{C}^N$ with the span of a pure k-form in $\Lambda\mathbb{C}^N$. The distinction between these two objects is unimportant on the level of bundles since they are isomorphic, but in calculations, it is occasionally more convenient to refer to one or the other. With a slight abuse of notation, we shall refer to both bundles as $\mathcal{E}(\tilde{K})$, although the line bundle is usually called the *determinant bundle* $\Lambda^k\mathcal{E}(\tilde{K})$ associated to the k-plane bundle $\mathcal{E}(\tilde{K})$.

In the context of the linearized Gray-Scott system, let $\tilde{K} \subset \mathcal{R}_+$ be a simple closed contour such that the gap condition (2.9) is satisfied for $\tilde{\lambda} \in \tilde{K}$, let $\xi = \xi(\tau)$ be the change of variables in (3.18), and define

$$\begin{aligned} \alpha(\tau, \tilde{\lambda}, \delta) &= e^{-(\mu_1+\mu_2)\xi(\tau)}\eta(\xi(\tau), \tilde{\lambda}, \delta) \\ \beta(\tilde{\lambda}, \delta) &= e_1(\tilde{\lambda}) \wedge e_2(\tilde{\lambda}, \delta) \end{aligned} \tag{2.15}$$

where $\eta(\xi, \tilde{\lambda}, \delta)$ is the 2-form in (2.12).

It follows from the defining condition (2.12) of η at $-\infty$ that $\beta(\tilde{\lambda}, \delta)$ extends $\alpha(\tau, \tilde{\lambda}, \delta)$ continuously to B_- at $\tau = -1$. In order to characterize the behavior of $\eta(\xi, \tilde{\lambda}, \delta)$ for large ξ we show that there is an exponential estimate for η as $\xi \to +\infty$ of the form

$$|e^{-(\mu_1+\mu_2)\xi}\eta(\xi, \tilde{\lambda}, \delta)| \leq Ke^{\max\{-\Gamma(\tilde{\lambda},\delta),0\}\xi}, \tag{2.16}$$

uniformly for $\tilde{\lambda} \in \tilde{K}$, where $\Gamma(\tilde{\lambda}, \delta)$ is the spectral gap (see (2.9)). This is because η satisfies an associated linear system $\eta' = a^{(2)}(\xi, \tilde{\lambda}, \delta)\eta$ whose asymptotic matrix $a_0^{(2)}(\tilde{\lambda}, \delta)$ has eigenvalues $\mu_i + \mu_j$ with $i \neq j$. The maximal growth rate of η is therefore determined by the eigenvalue of $a_0^{(2)}(\tilde{\lambda}, \delta)$ with largest real part. This eigenvalue is $\mu_1 + \mu_i$ where $i = 2$ if $\Gamma > 0$ and $i = 3$ if $\Gamma < 0$. This statement implies (2.16).

Let ξ_+ be defined by $\xi_+ = \xi(\tau_+)$, so that ξ_+ is large and positive. It then follows from (2.14) and (2.16) that

$$\begin{aligned} \alpha(\tau_+, \tilde{\lambda}, \delta) &= g(\tilde{\lambda}, \delta) e_1(\tilde{\lambda}) \wedge e_2(\tilde{\lambda}, \delta) + \gamma(\tilde{\lambda}, \delta) \\ |g|, |b| &= \mathbf{O}(e^{|\Gamma(\tilde{\lambda}, \delta)|\xi_+}). \end{aligned} \tag{2.17}$$

Here, $\gamma(\tilde{\lambda})$ is in the subspace of $\Lambda^2\mathbb{C}^4$ spanned by $e_i \wedge e_j$ with $(i,j) \neq (1,2)$. Now let $\zeta(\xi, \tilde{\lambda}, \delta)$ be the 2-form in (2.12) with defining condition at $\xi = +\infty$. Since $\zeta(\tilde{\lambda}, \delta) = e_3 \wedge e_4$ it therefore follows that $\zeta(\tilde{\lambda}, \delta) \wedge \gamma(\tilde{\lambda}) = 0$. This condition uniquely determines clutching function $g(\tilde{\lambda})$, and hence, the identification map F for the bundle.

PROPOSITION 2.8. *Let $\tilde{K}$ be a simple closed contour such that (2.9) is satisfied along $\tilde{K}$. Let $\xi_+ = \xi_+(\delta) = \delta^{-s}$ for some $s > 0$. Suppose also that $g(\tilde{\lambda}, \delta)$ in (2.17) is uniformly bounded away from zero for all small $\delta > 0$ and for $\tilde{\lambda} \in \tilde{K}$. Let $\mathcal{E}(\tilde{K}, \delta)$ be the bundle constructed above by the clutching function $g = g(\tilde{\lambda}, \delta)$. Then $c_1(\mathcal{E}(\tilde{K}, \delta)) = W(D(\tilde{K}, \delta))$, where $D(\tilde{\lambda}, \delta)$ is the analytic continuation of the Evans function in (2.13).*

PROOF. Evaluating D at $\xi = \xi_+$, to obtain

$$D(\tilde{\lambda}, \delta) = \eta(\xi_+, \tilde{\lambda}, \delta) \wedge \zeta(\xi_+, \tilde{\lambda}, \delta).$$

Use (2.12), (2.15), and (2.17) in the above together with the fact that the trace of a_o is zero to obtain

$$D(\tilde{\lambda}, \delta) = [g(\tilde{\lambda}, \delta) e_1(\tilde{\lambda}) \wedge e_2(\tilde{\lambda}, \delta) + \gamma(\tilde{\lambda})] \wedge [e_3(\tilde{\lambda}, \delta) \wedge e_4(\tilde{\lambda}) + \mathbf{O}(-\tilde{\kappa}\xi_+)].$$

Now note that $\gamma(\tilde{\lambda}) \wedge e_3(\tilde{\lambda}, \delta) \wedge e_4(\tilde{\lambda}) = 0$. It then follows from the estimate for g in (2.17) that

$$D(\tilde{\lambda}, \delta) = g(\tilde{\lambda}, \delta)[e_1 \wedge e_2 \wedge e_3 \wedge e_4] + \mathbf{O}(e^{(-\tilde{\kappa} + |\Gamma(\tilde{\lambda}, \delta)|)\xi_+}).$$

Since $\Gamma = \mathbf{O}(\delta)$, we have that $-\tilde{\kappa} + |\Gamma(\tilde{\lambda}, \delta)| < 0$ uniformly along $\tilde{K}$. The error term is therefore exponentially small in δ, since $\xi_+(\delta) = \delta^{-s}$ with $s > 0$. From Lemma 2.1 we see that the wedge of the four eigenvectors of $a_0(\tilde{\lambda}, \delta)$ is of order δ^β, and hence this wedge is at worst algebraically small. Finally, our assumption that $g(\tilde{K})$ is uniformly bounded away from the origin and the above together with the argument principle imply that $W(D(\tilde{K}, \delta)) = W(g(\tilde{K}, \delta))$.

It then follows that both $g(\tilde{K}, \delta)$ and $D(\tilde{K}, \delta)$ are homotopic to one another in the punctured complex plane. The connection with this homotopy type and first Chern number of the bundle then follows as in [**1**]. □

REMARK 2.9. Later in the paper, we shall verify the crucial hypothesis concerning the nonvanishing of g along various contours $\tilde{K}$. This is the main estimate of the paper.

2.3.1. A Winding Number Formula. It will be convenient to complete this discussion by recalling a standard winding number formula from complex variables. Let $\mathcal{E}(K) = \mathcal{E}_+ \cup_f \mathcal{E}_-$ be a bundle constructed as in the previous section with clutching function $g(\lambda)$. We emphasize that g is only known to be analytic on K, and that it may have poles and other singularities inside K. The following theorem can be used either when $K \subset S_1$ is a contour in the standard complex plane, or when $K = \tilde{K} = \partial\tilde{N}$ is a contour about the branch point $\lambda_*(\delta)$ on the Riemann surface.

THEOREM 2.10. *In local coordinates, write $\tilde{\lambda} \in \mathcal{R}$ as $(\lambda, \pm\sqrt{\lambda - \lambda_*(\delta)})$, where $\sqrt{\ }$ is the standard branch of the square root function with branch cut along the negative real axis and where $\lambda \in \mathbb{C}$. Suppose that $g(\tilde{\lambda})$ can be expressed locally as*

$$g(\tilde{\lambda}) = \begin{cases} G(\sqrt{\lambda - \lambda_*(\delta)}) & \text{for } \tilde{\lambda} \in S_1 \\ G(-\sqrt{\lambda - \lambda_*(\delta)}) & \text{for } \tilde{\lambda} \in S_2, \end{cases}$$

where S_1, S_2 are the two sheets of the Riemann surface $\mathcal{R}$ and $G = G(\zeta)$ is a meromorphic function of ζ with $G(\zeta) = \sum_{j=\ell}^{\infty} a_j \zeta^j$ and $a_\ell \neq 0$.

(i) Let $K \subset S_1$ be a simple closed contour for which the bundle $\mathcal{E}(K)$ can be constructed with clutching function g. Then

$$c_1(\mathcal{E}(K, \delta)\ ) = W(g(K)) = Z - P$$

where Z is the multiplicity of roots of g in K^{int}, P is the multiplicity of poles of g in K^{int}.

(ii) Suppose that $K = \tilde{K} = \partial\tilde{N}$ is a neighborhood of the branch point $\lambda_*(\delta)$ on $\mathcal{R}$. Then

$$c_1(\mathcal{E}(\tilde{K}, \delta)\ ) = W(g(\tilde{K})) = Z - P + \ell,$$

where Z and P are the multiplicity of zeros and poles of g inside $\tilde{N} \setminus \{\lambda_*(\delta)\}$.

PROOF. The proof of (i) is a standard formula from elementary complex variables. In order to prove (ii), we use the same argument as in (i) to localize the line integral

$$W(g(\tilde{\lambda})) = \frac{1}{2\pi i} \int_{\tilde{K}} \frac{g'(s)}{g(s)}\, ds,$$

about contours winding about each of the poles and roots of g inside $\tilde{N}$. However, we also need to introduce an additional simple closed contour $\tilde{\Lambda} \subset \tilde{K}^{int}$ such that the only singularity or root of g inside $\tilde{\Lambda}$ is the branch point at $\lambda_*(\delta)$; let Λ_1 be the projection of $\tilde{\Lambda}$ on S_1. We then have that

$$\begin{aligned} W(g(\tilde{\Lambda})) &= \frac{1}{2\pi i} \int_{\tilde{\Lambda}} \frac{g'(\tilde{\lambda})}{g(\tilde{\lambda})}\, d\tilde{\lambda} \\ &= \frac{1}{2\pi i} \int_{\Lambda_1} \frac{\frac{d}{d\lambda} G(\sqrt{\lambda - \lambda_*(\delta)})}{G(\sqrt{\lambda - \lambda_*(\delta)})}\, d\lambda + \frac{1}{2\pi i} \int_{\Lambda_1} \frac{\frac{d}{d\lambda} G(-\sqrt{\lambda - \lambda_*(\delta)})}{G(-\sqrt{\lambda - \lambda_*(\delta)})}\, d\lambda \\ &= \frac{1}{2\pi i} \int_{\Lambda_1} [\frac{\ell}{2}(\lambda - \lambda_*(\delta))^{-1} + \mathbf{O}(1)]\, d\lambda \\ &\quad + \frac{1}{2\pi i} \int_{\Lambda_1} [\frac{\ell}{2}(\lambda - \lambda_*(\delta))^{-1} + \mathbf{O}(1)]\, d\lambda \\ &= \ell. \end{aligned}$$

□

2.4. Outline and Endgame of the Proof

The outline of the proof generally follows along the lines developed in [**13**], in which two distinguished independent solutions y_1 and y_2 in the set $\Phi(\xi, \lambda, \delta)$ are tracked as they evolve over the underlying wave. Here, y_2 is distinct from the solution y_2^* used in the definition of the Evans function. The eventual goal is to determine conditions so that $\lambda \in K$ is *not* an eigenvalue of the perturbed wave, so that the bundle $\mathcal{E}(K, \delta)$ can be formed in the manner described in the previous section. This is achieved by using y_1, y_2 to build one-dimensional summands of the 2-plane bundle $\mathcal{E}(K, \delta)$, and by relating the solutions in the fibers of these line bundles to certain reduced eigenvalue problems. In fact, the reduced eigenvalue problems provide the crucial estimates of the solutions to the perturbed equations. The eventual condition for the tracking of the perturbed solutions and for the construction of their associated bundles is that K should be disjoint from the union of the eigenvalues of the two reduced eigenvalue problems, the linearization (1.14) about the scalar Fisher wave (for y_1) and the NLEP (for y_2).

In order to form the bundles associated to these two solutions, we need to track their span in $\mathbb{C}^4$ as they evolve over the wave. To this end, we follow their projective images $\hat{y}_i(\xi, \tilde{\lambda}, \delta) \in \mathbb{CP}^3$, where $\mathbb{CP}^3$ is the space of complex lines in $\mathbb{C}^4$. The linear system (2.4) induces a system of differential equations on projective space which, in local coordinates, forms a generalized, nonautonomous quadratic system which generalizes the familiar Riccati equation from Sturm-Liouville theory. We denote this system by

$$\hat{y}' = \hat{a}(\xi, \tilde{\lambda}, \delta; \hat{y}), \tag{2.18}$$

where $\hat{a}$ is the vector field on the tangent space to $\mathbb{CP}^3$ induced by the matrix a in (2.4). These equations are the *projectivized system* associated to (2.4).

Notational conventions. (i) We shall frequently use (2.18) to estimate solutions of the associated linear equations (2.4), and vice versa. With a slight abuse of notation, we shall refer to $y \in \hat{y}$ as a nonzero vector in the line in $\mathbb{C}^4$ represented by $\hat{y} \in \mathbb{CP}^3$.

(ii) Since $\mathbb{CP}^3$ is a compact manifold, it is convenient to have a measure of distance in $\mathbb{CP}^3$ in terms of a fixed, globally defined metric $\rho[\ ,\ ]$. Convergence and approximation results regarding solutions of (2.18) will be stated in terms of this metric.

The proof proceeds by first tracking the solution $\hat{y}_1(\xi, \tilde{\lambda}, \delta)$ of (2.18) as $\xi \to +\infty$, where $y_1 \in \hat{y}_1$ is the solution of Lemma 2.2 used to form the Evans function. To this end, we construct in Chapter 3 a positively invariant tube $\Omega_1 \subset \mathbb{CP}^3 \times \mathbb{R}$ containing $(\hat{y}_1, \xi)$, whenever $\tilde{\lambda} = (\lambda, \pm\sqrt{(\lambda - \lambda_*(\delta))})$ and $\lambda \in S_1$ is *not* an eigenvalue of the linearization (1.14) about the Fisher wave. The structure of this tube implies the limit $\hat{y}_1 \to \hat{e}_1$ as $\xi \to +\infty$, and hence, the existence of a fast bundle $\mathcal{E}_f(\tilde{K}, \delta)$ whenever the projection of $\tilde{K}$ on S_1 is disjoint from the spectrum of (1.14). This approximation enables us to calculate the index of the fast bundle by continuing it to a reduced bundle $\mathcal{E}_{fisher}(\tilde{K})$ constructed directly from (1.14).

The tracking of $\hat{y}_1$ also provides key information needed in defining and tracking a new, uniquely determined *slow* solution $\hat{y}_2$ which remains near the slow subspace spanned by e_2, e_3 over the slow manifolds. The main estimates of this paper are to track $\hat{y}_2$ and also, an associated solution $y_2 \in \hat{y}_2$ of (2.4) over the fast field. The tracking of these solutions can then be used to form a slow bundle $\mathcal{E}_s(\tilde{K}, \delta)$ using the clutching operation described in Chapter 2, section 3. In particular, the clutching function for this bundle has roots which, to leading order, coincide with the roots of the NLEP (1.13). Hence, if $\tilde{K}$ is disjoint from the spectrum of the Fisher wave and also, the roots and poles of the NLEP, the slow bundle can be formed. This analysis is carried out in Chapter 4.

It is here that the choice of the solution y_2 used in the construction of the bundle differs from the choice of the solution y_2^* used in the construction of the Evans function. The corresponding solution y_2 of the linear equations may have poles interior to $\tilde{K}$, whereas y_2^* must be analytic on $\tilde{K} \cup \tilde{K}^{int}$.

The full bundle $\mathcal{E}(K, \delta)$ is then the Whitney sum $\mathcal{E}(K, \delta) = \mathcal{E}_f(K, \delta) \oplus \mathcal{E}_s(K, \delta)$, and the index calculation for the full bundle follows from the additivity property of c_1

$$c_1(\mathcal{E}(K, \delta)) = c_1(\mathcal{E}_f(K, \delta)) + c_1(\mathcal{E}_s(K, \delta)) \tag{2.19}$$

on Whitney sums. The various statements in Theorem 1.3 are verified in Chapter 5 by using the error estimates of Chapters 3 and 4 to continue the fast and slow summands of the perturbed wave to bundles generated by the unperturbed Fisher wave and the NLEP, respectively. In each of these singular limits, the calculation reduces to an algebraic count of the roots and poles of the clutching functions for each of the summands interior to various contours $\tilde{K}$.

CHAPTER 3

Tracking the Fast Subbundle

In this chapter, we present two estimates for the fast projectivized solution $\hat{y}_1(\xi, \tilde{\lambda}, \delta)$ of (2.18), where $y_1 \in \hat{y}_1$ is as in Lemma 2.2. A "large λ" result is proved in Chapter 3, section 2 for $|\lambda| \geq H$, for a large, **O**(1) constant H. A second result for $\tilde{\lambda}$ in bounded regions of $\mathcal{R}_+$ is presented in Chapter 3, section 3.

The tracking of the fast and slow subbundles depends critically on the structure of the fundamental one-pulse solution $(u_o(\xi;\delta), v_o(\xi;\delta))$. The next lemma provides some estimates that are necessary for the forthcoming analysis.

LEMMA 3.1.
(Fast estimate) Let $|\xi| < 1/\delta^r$ for $0 \leq r < 1-\beta$, then there exist $O(1)$ constants C_1 and C_2 such that

$$|u_o(\xi) - \bar{c}| < C_1\delta^{1-\beta}, \quad |v_o(\xi) - S(\xi)| < C_2\delta^{1-\beta}, \tag{3.1}$$

where $\bar{c}$ and $S(\xi)$ are defined in (1.7).
(Slow estimate) Let $|\xi| > 1/\delta^s$ for $s > 0$, then there exists an $O(1)$ constant C_3 such that

$$|u_o(\xi)v_o(\xi)| < C_3 e^{-\sqrt{b}/\delta^s}/\delta^{3\beta/2}. \tag{3.2}$$

(Essential estimate) Given any $\rho > 0$, there exist $T(\rho), \delta(\rho) > 0$ such that if $|\xi| \geq T(\rho)$ and $\delta < \delta(\rho)$, then

$$2u_o(\xi,\delta)v_o(\xi,\delta) < \rho \qquad \text{and} \qquad v_o(\xi,\delta)^2 < \rho.$$

PROOF. The (rescaled) pulse solution $v_o(\xi,\delta)$ is bounded for all ξ (by construction). Therefore, it is a straightforward calculation in (1.6) to show that the error $\tilde{u}(\xi)$ defined by $u_o(\xi,\delta) - \bar{c} = \delta^{1-\beta}\tilde{u}(\xi)$ remains bounded for $|\xi| < 1/\delta^r$. The estimate on $v_o(\xi,\delta)$ in the fast field then follows directly from the Fenichel invariant fiber theorem [**10**] applied to the (rescaled) 4-dimensional system of differential equations for $(u_o, u_{o\xi}, v_o, v_{o\xi})$ (1.6), since the estimate on $u_o(\xi)$ controls the behavior of the base points of the fiber. Moreover, the exponential contraction in forward (and backward) 'time' along the stable (and unstable) fibers yields that v_o must remain exponentially close to the slow manifold $\mathcal{M} = \{(u_o, u_{o\xi}, v_o, v_{o\xi}) | v_o = v_{o\xi} = 0\}$ of (1.6) for $|\xi| > 1/\delta^s$:

$$|v_o(\xi,\delta)| < Ce^{-\sqrt{b}|\xi|}.$$

Estimate (3.2) then follows from the rescaling (1.5), since the unscaled solution $U_o(x,\delta)$ satisfies $0 < U_o \leq 1$ by construction [**8**]. The essential estimate follows immediately from a combination of (3.1) and (3.2). □

3.1. Asymptotic and Reduced Equations in the Fast Scaling

We shall use various limiting systems to control the perturbed eigenvalue problem. These are the asymptotic equations at $|\xi| = \infty$, and the (fast) reduced equations at $\delta = 0$. We record these equations and their projectivizations in suitable local coordinates here for future reference.

The fast reduced eigenvalue problem is obtained by letting δ (and, hence also ε) $\to 0$ in (2.4), (2.5). In the limit as $|\xi| \to \infty$, this further reduces to

$$y' = a_R(\lambda)y, \tag{3.3}$$

where the coefficient matrix is

$$a_R(\lambda) = \begin{pmatrix} 0 & 0 & 0 & 0 \\ 0 & 0 & 0 & 0 \\ 0 & 0 & 0 & 1 \\ 0 & 0 & b(1+\lambda) & 0 \end{pmatrix}.$$

It then follows from Lemma 3.1 that

$$\|a_R(\lambda) - a(\xi, \lambda, \delta)\| < \rho \tag{3.4}$$

for $|\xi| \geq T(\rho)$ and $\delta < \delta(\rho)$. The eigenvalues and eigenvectors of $a_R(\lambda)$ are given by

$$\mu_{1R} = \sqrt{b(1+\lambda)}; \quad \mu_{4R} = -\sqrt{b(1+\lambda)}$$
$$\mu_{2R} = 0; \quad \mu_{3R} = 0,$$

and

$$e_{1R} = (0,0,1,\mu_{1R})^t, \quad e_{4R} = (0,0,1,\mu_{4R})^t, \quad e_{2R} = (1,0,0,0)^t, \quad e_{3R} = (0,1,0,0)^t.$$

It is also useful later to note that

$$\text{Re}\ \mu_{4R}(\lambda) < 0 < \text{Re}\ \mu_{1R}(\lambda) \tag{3.5}$$

uniformly for all $\lambda \in B$ (see (1.15)).

Recall the projectivized equations (2.18) associated with (2.4) which were introduced in Chapter 2, section 4. In a similar fashion, the linear system (3.3) also induces projectivized equations on $\mathbb{CP}^3$:

$$\hat{y}' = \hat{a}_R(\lambda; \hat{y}). \tag{3.6}$$

This projectivized system has rest points at $\hat{e}_{iR}(\lambda)$ for $i = 1–4$, since $a_R(\lambda)e_{iR}(\lambda) = \mu_R(\lambda)e_{iR}(\lambda)$ for each i. Moreover, it is easily seen that the spectrum of the projectivized vector field linearized at $\hat{e}_{jR}(\lambda)$ is given by the differences of the remaining eigenvalues and $\mu_{jR}(\lambda)$. This is established by an explicit calculation of the Jacobian of $\hat{a}_R(\lambda)$ at the rest points, in local coordinates (see e.g. (3.10) below). In particular, linearizing about $\hat{e}_{1R}(\lambda)$ yields: $\sigma(d\hat{a}_R(\lambda, \hat{e}_{1R}(\lambda))) = \mu_{iR}(\lambda) - \mu_{1R}(\lambda)$ for $i = 2, 3, 4$. Now, by (3.5), there exists an $\tilde{\alpha} > 0$ such that $\text{Re}\,[\mu_{iR}(\lambda) - \mu_{1R}(\lambda)] < -\tilde{\alpha}$ for $i = 2, 3, 4$ and $\text{Re}\lambda \geq -1/2$. Hence, $\hat{e}_{1R}(\lambda)$ is an attracting rest point of (3.6) when $\lambda \in B$, with a uniform exponential rate of attraction $-\tilde{\alpha}$. (As noted earlier in Chapter 1, the third eigenvalue of the Fisher wave occurs at $\lambda = -3/4$. The restriction to $\lambda \in B$ is imposed in order to avoid this eigenvalue.)

In Chapter 3, section 4, we will also need to study the fast reduced limit at $\delta = 0$ for finite ξ by setting δ and hence ε to zero in (2.4). The resulting equations are

$$
\begin{aligned}
u' &= 0 \\
p' &= 0 \\
v' &= q \\
q' &= -S(\xi)^2 u + \gamma_{vR}(\xi, \lambda) v,
\end{aligned} \tag{3.7}
$$

where $\gamma_{vR}(\xi, \lambda) = -2\bar{c}S(\xi) + b(1 + \lambda)$ (recall (1.7)). The projectivized version of (3.7) is

$$
\hat{y}' = \hat{a}(\xi, \lambda, 0, \hat{y}). \tag{3.8}
$$

Near $\hat{e}_{1R}(\lambda)$ on $\mathbb{C}\mathbf{P}^3$, local coordinates may be defined by taking $z \equiv (\frac{u}{v}, \frac{p}{v}, \frac{q}{v}) \in \mathbb{C}^3$. By (3.5), this is a good coordinate system when $\lambda \in B$. In terms of this local coordinate, the projectivized equations (2.18), (3.8) and (3.6) are

$$
\begin{aligned}
z' &= F(\xi, \lambda, \delta, z) \\
z' &= F_R(\xi, \lambda, z) \\
z' &= F_R(\lambda, z),
\end{aligned} \tag{3.9}
$$

respectively. An explicit determination of $F(\xi, \lambda, \delta, z)$ and $F_R(\xi, \lambda, z)$ yields:

$$
\begin{aligned}
z_1' &= \varepsilon\, z_2 - z_1 z_3 & z_1' &= -z_1 z_3 \\
z_2' &= \varepsilon \gamma_u z_1 + 2\varepsilon u_o v_o - z_2 z_3 & z_2' &= -z_2 z_3 \\
z_3' &= -v_o^2 z_1 + \gamma_v - z_3^2 & z_3' &= -S(\xi)^2 z_1 + \gamma_{vR} - z_3^2.
\end{aligned} \tag{3.10}
$$

The local vector field for $F_R(\lambda, z)$ is found by setting $S(\xi) = 0$ and $\gamma_{vR} = b(1 + \lambda)$ in the above system for $F_R(\xi, \lambda, z)$.

The results stated above for the projectivized systems (2.18) and (3.6) directly imply that the rest point point $z_{1R}(\lambda) = (0, 0, \mu_{1R}(\lambda))$ identified with $\hat{e}_{1R}(\lambda)$ is an attracting rest point of the third system in (3.9). In fact, let $B_R(\lambda) \equiv dF_R(z_{1R}(\lambda), \lambda)$ so that $\sigma B_R(\lambda) = \{\mu_{iR}(\lambda) - \mu_{1R}(\lambda) : i = 2, 3, 4\}$. Then, one observes from Lemma 2.1 that $B_R(\lambda)$ has strictly stable spectrum for $\lambda \in B$.

3.2. A Large λ Estimate

In the ensuing analysis, it will be important to obtain estimates that are uniform for all $\lambda \in B$. Since $|\lambda|$ can be arbitrarily large in this region, we present an estimate for the solution $\hat{y}_1$ of (2.18) for large λ. This is complicated by the fact that the term $\varepsilon\gamma_u$ in the second equation in (3.10) is then of indeterminate size due to the presence of λ in γ_u.

To this end, it will be convenient to introduce a second small parameter $d = |b(1+\lambda)|^{-1/2}$. When $\lambda \in B$, we have that $b(1+\lambda) = d^{-2}e^{i\theta}$ with $|\theta| \leq \pi/2 + \mathbf{O}(d^2)$. Let $\omega = \omega(\theta) = e^{i\theta/2}$ so that Re $\omega > \sqrt{2}/2 - kd^2$. Rescale the local coordinate z defined above by setting $\zeta_i(t) = z_i(dt)$ for $i = 1, 2$, and $\zeta_3(t) = dz_3(dt)$. The equations for ζ are

$$
\begin{aligned}
\zeta_1' &= \delta^{1-\beta} d\zeta_2 - \zeta_1\zeta_3 \\
\zeta_2' &= (\frac{\delta^{1+\beta}}{d}\omega^2 - db\delta^{1+\beta} + d\delta^{1-\beta}(v_o^2 + a\delta^{2+\beta}))\zeta_1 + 2\delta^{1-\beta} d u_o v_o - \zeta_2\zeta_3 \\
\zeta_3' &= \omega^2 - d^2(v_o^2\zeta_1 + 2u_o v_o) - \zeta_3^2.
\end{aligned}
$$

We define a second change of coordinates by setting $\rho(t) = \zeta_1' = \delta^{1-\beta} d\zeta_2 - \zeta_1\zeta_3$ and $\gamma(t) = \zeta_3 - \omega$. The equations in the local coordinates (ζ_1, ρ, γ) are then

$$\begin{aligned} (3.11)\quad \zeta_1' &= \rho \\ \rho' &= -\omega^2\zeta_1 - 2\omega\rho - 2\rho\gamma + \mathbf{O}(\delta^2 + d^2)\zeta_1 + \mathbf{O}(d^2\zeta_1^2) + \mathbf{O}(\delta^{2(1-\beta}d^2) \\ \gamma' &= -2\omega\gamma - \gamma^2 + \mathbf{O}(d^2) + \mathbf{O}(d^2)\zeta_1. \end{aligned}$$

LEMMA 3.2. *(i) There exists $d_o > 0$ and $\delta_o > 0$ such that for all $\omega = e^{i\theta/2}$ with $|\theta| \le \pi/2 + \mathbf{O}(d^2)$, there exists an $\mathbf{O}(1)$ neighborhood $V_o(\omega)$ of the origin of the (ζ_1, ρ, γ) plane such that $V_o(\omega)$ is positively invariant for (3.11) for all $\delta < \delta_o$ and $d < d_o$.*

(ii) Let $z(\xi, \lambda, \delta)$ be the local coordinate representation of the solution $\hat{y}_1(\xi, \lambda, \delta)$ of (2.18), where $y_1 \in \hat{y}_1$ is the solution of (2.4) of Lemma 2.2. Let $b(1+\lambda) = d^{-2}\omega^2$ where $|\omega| = 1$, and define

$$\Gamma(t, \lambda, \delta) = (\zeta_1, \rho, \gamma)(t, \lambda, \delta) = (z_1, \varepsilon dz_2 - dz_1 z_3, dz_3 - \omega)(dt, \lambda, \delta).$$

Let $H = \frac{bd_o^2+1}{bd_o^2}$, and suppose that $\lambda \in B \cap \{|\lambda| > H\}$ (see (1.16)). Then $\Gamma(t, \lambda, \delta) \in V_o(\omega)$ for all $t \in \mathbb{R}$ and $\lim_{\xi\to+\infty} \hat{y}_1(\xi, \lambda, \delta) = \hat{e}_1(\lambda)$ for all $\delta < \delta_o$.

PROOF. The system (3.11) is of the form $\Gamma' = A(\omega)\Gamma + \mathbf{O}(\delta^2 + d^2)(1 + |\Gamma|) + \mathbf{O}(|\Gamma|^2)$, where

$$A(\omega) = \begin{pmatrix} 0 & 1 & 0 \\ -\omega^2 & -2\omega & 0 \\ 0 & 0 & -2\omega \end{pmatrix}.$$

Note that all the $\mathbf{O}$ terms are uniform for all small δ and d. Even though $A(\omega)$ is not diagonalizable, it has eigenvalues $-\omega, -\omega, -2\omega$ with strictly negative real part since $\mathrm{Re}\,\omega > \sqrt{2}/2$. Note that this can be achieved by an $\mathbf{O}(d^2)$ correction in ω. This correction is of the same magnitude as the other higher order terms in the equations fro Γ.but $\mathbf{O}(1)$ neighborhood N of the origin such that the perturbed wave has Set $L = \Lambda(\omega)\Gamma$ where $\Lambda(\omega)$ is the 3×3 matrix of eigenvectors and generalized eigenvectors of $A(\omega)$. We therefore have that

$$L' = J(\omega)L + \mathbf{O}(\delta^2 + d^2)(1 + |L|) + \mathbf{O}(|L|^2), \tag{3.12}$$

where

$$J(\omega) = \begin{pmatrix} -\omega & r & 0 \\ 0 & -\omega & 0 \\ 0 & 0 & -2\omega \end{pmatrix}.$$

The matrix $\Lambda(\omega)$ can be selected so that the off-diagonal entry r is sufficiently small (relative to ω) so that $\mathrm{Re}\,\bar{L}^t JL < -(\sqrt{2}/4)|L|^2$. Let $g(t) = (\bar{L}^t L)^{1/2}$, so that from the above,

$$g' < -\frac{\sqrt{2}}{4}|L| + K|L|^2 + K(\delta^2 + d^2)(1 + g)$$

for some $\mathbf{O}(1)$ constant $K > 0$. Let $0 < \alpha < \sqrt{2}/4$ be fixed, and let $d_o, \delta_o > 0$ be such that $-\sqrt{2}/4 + K(d_o^2 + \delta_o^2) < -\alpha$. It then follows that $g' \le Kg^2 - \alpha g + K(d_o^2 + \delta_o^2)$. By decreasing δ_o and d_o further if necessary, select δ_o and d_o so that $d_o^2 + \delta_o^2 < \alpha^2/(4K^2)$. Now set $\eta = \alpha/2K$. It then follows that if $g = \eta$ then $g' < 0$, so that the ball $B_\eta = \{L : |L| < \eta\}$ is positively invariant for (3.12). The ball $V_o(\omega) = \Lambda^{-1}(\omega)B_\eta$ then determines a positively invariant set for (3.11).

To prove (ii), note first that the conditions $|\lambda| > H$ and $\lambda \in B$ imply that $d < d_o$ and $\omega = e^{i\theta}$ with $|\theta| < \pi/2$. Next note that by its defining condition at $-\infty$, the solution $\hat{y}_1$ lies arbitrarily close to $\hat{e}_1(\lambda)$ for all sufficiently negative ξ. The image of $\hat{e}_1(\lambda)$ in the local coordinate z is $(0, 0, \sqrt{b(1+\lambda)})$. This point is then mapped into $(0, 0, \omega)$ in the ζ system, and then into the origin of the Γ system. Hence if $\Gamma(t)$ is as in (ii) of the Lemma, then $\Gamma(t) \in V_o(\omega)$ for all sufficiently negative t, so that by (i), $\Gamma(t) \in V_o(\omega)$ for all $t \in \mathbb{R}$. The origin is an attractor for the asymptotic system associated to (3.12) at $+\infty$, and from the manner in which η was selected above, the ball B_η is in the domain of attraction of the origin for the asymptotic flow. It then follows from standard ω-limit set theorems that $\Gamma(t)$ tends to the origin as $t \to +\infty$. □

3.3. Invariant Sets for the Projectivized Flow

We next consider the problem of tracking $\hat{y}_1(\xi, \tilde{\lambda}, \delta)$ for bounded $\tilde{\lambda} \in \mathcal{R}_+$. To this end we construct a positively invariant tube $\Omega_1 \subset \mathbb{CP}^3 \times \mathbb{R}$ that constrains the unique solution $\hat{y}_1(\xi, \tilde{\lambda}, \delta)$ that is backward asymptotic to $\hat{e}_1(\tilde{\lambda})$ as $\xi \to -\infty$ to be also forward asymptotic to $\hat{e}_1(\tilde{\lambda})$ for all λ on a bounded subset of $\mathcal{R}_+$ that does not intersect the spectrum of the linearized Fisher equation (1.14). This sort of tracking lemma arises frequently in linearized singular perturbation problems. The "Elephant Trunk Lemma" is a general method that has been developed in connection with such problems. This technique was initiated by Jones [**20**] in his study of the FitzHugh-Nagumo system, and it was generalized and applied in a variety of other applications (see e.g. [**12**]). In particular, the structure of the problem at hand most closely resembles the application in [**12**]. We freely refer to this paper in verifying the technical details that are common to the two problems.

The structure of Ω_1 reflects the singular structure of the underlying wave. It is the union of three tubes, two of which track $\hat{y}_1$ when the underlying wave is in the left and right slow manifolds. These intervals are $\xi < T_L$ and $\xi > T_R$, respectively, where T_L and T_R are large but $\mathbf{O}(1)$ negative and positive constants, respectively. The third tube tracks $\hat{y}_1$ over the fast field, which for this solution is defined to be the interval $[T_L, T_R]$.

We first introduce certain quantities $T(r, H)$ and $\delta(r, H)$ which are used in the definition of T_L, T_R given later. Here $|\lambda| \leq H$, where H is the $\mathbf{O}(1)$ parameter determined in Lemma 3.2, and r is another given positive parameter. For given H and r there exist $T(r, H) > 0$ and $\delta(r, H) > 0$ such that

$$\|F_R(z, \tilde{\lambda}) - F(\xi, \tilde{\lambda}, \delta, z)\| < r \tag{3.13}$$

for all $|\xi| \geq T(r, H)$ and $\delta \leq \delta(r, H)$. This estimate follows directly from the estimate (3.4).

Definition 3.3. A subset $V \times I \subset \mathbb{CP}^3 \times \mathbb{R}$ is positively invariant relative to I for the autonomous system

$$\begin{aligned} z' &= F(\xi, \tilde{\lambda}, \delta, z) \\ \xi' &= 1 \end{aligned} \tag{3.14}$$

if $(\hat{y}_0, \xi_0) \in V \times I$ implies $(\hat{y}(\xi), \xi) \in V \times I$ for all $\xi \geq \xi_0$ and $\xi \in I$, where $\hat{y}(\xi_0) = \hat{y}_0$.

We look for relatively invariant sets of (3.14) using neighborhoods of the attracting fixed point $z_{1R}(\lambda)$. Let $V_0(\tilde{\lambda}) \subset C^3$ be an attracting neighborhood of $z_{1R}(\tilde{\lambda})$

such that for all points p on the boundary $\partial V_0(\tilde{\lambda})$ there exists an α independent of p and $\tilde{\lambda} \in \mathcal{R}_+$ for which

$$\mathbf{n}(p) \cdot F_R(p, \tilde{\lambda}) < -\alpha, \tag{3.15}$$

where $\mathbf{n}(p)$ denotes the outward unit normal to $\partial V_0(\tilde{\lambda})$ at p.

LEMMA 3.4. *Let $\delta_0 = \delta(\frac{\alpha}{2}, H)$ and $T = T(\frac{\alpha}{2}, H)$ be such that (3.13) holds with $r = \alpha/2$, for $|\xi| \geq T$ and for $\delta < \delta_0$. Then, for $\delta < \delta_0$ and $|\xi| \geq T$, the sets $V_0(\tilde{\lambda}) \times (-\infty, -T]$ and $V_0(\tilde{\lambda}) \times [T, \infty)$ are positively invariant for (3.14) relative to $(-\infty, -T]$ and $[T, \infty)$, respectively.*

PROOF. The hypotheses of the lemma and (3.15) imply $\mathbf{n}(p) \cdot F(\xi, \tilde{\lambda}, \delta, p) < -\alpha/2$ for $\delta < \delta_0$ and $|\xi| \geq T$. □

Clearly, over the interval $-\infty < \xi \leq -T$, this lemma facilitates the tracking of the unique solution $\hat{y}_1(\xi, \tilde{\lambda}, \delta)$ of (2.18) that tends to $\hat{e}_1(\tilde{\lambda}, \delta)$ as $\xi \to -\infty$. Given that this solution lies inside the neighborhood $V_0(\tilde{\lambda})$ at $-\infty$, the lemma shows that it must stay inside the invariant set at least until $-T$. Hence, control over the solution's position during the interval $-\infty < \xi \leq -T$ is given by the size of the invariant tube.

We now turn to the tracking of this solution over the pulse. Clearly, the variables u and p of (3.7) are constant. Moreover, the choice of these constants determines the fast eigenvalue problem, and this choice is suggested by the projectivized version of (3.7). Namely, we seek a solution of (3.8) that connects $\hat{e}_{1R}(\tilde{\lambda})$ at $-\infty$ to $\hat{e}_{1R}(\tilde{\lambda})$ at $+\infty$. In local coordinates, $\hat{e}_{1R}(\tilde{\lambda}) = (0, 0, \mu_{1R}(\tilde{\lambda}))$, revealing that the u and p components of this connecting orbit vanish at $\pm\infty$. Hence, we set $u, p \equiv 0$ over the fast pulse, and the fast reduced eigenvalue problem becomes:

$$\begin{aligned} u &\equiv 0 \\ p &\equiv 0 \\ v' &= q \\ q' &= \gamma_{vR}(\xi, \lambda) v, \end{aligned} \tag{3.16}$$

which is simply the linearization (1.14) around the unperturbed Fisher wave. This Fisher wave has a simple unstable eigenvalue at $\lambda = \lambda_1 = 5/4$ and a simple eigenvalue at zero. These two are the only eigenvalues of (1.14) for

$$\tilde{\lambda} = (\lambda, \pm\sqrt{\lambda - \lambda_*(\delta)}) \in \mathcal{R}_+,$$

where $\mathcal{R}_+$ is as in (2.11).

Let $\tilde{N}$ be the small, $\mathbf{O}(1)$ neighborhood of $\lambda_*(\delta)$ defined above (2.11), let N be the projection of $\tilde{N}$ on the first sheet S_1 of $\mathcal{R}$. Thus $\tilde{K} = \partial\tilde{N}$ is a simple, closed contour on $\mathcal{R}$ which encloses both the branch point and the origin $\lambda = 0$ on S_1 in its interior. Finally, let U_1 be a small $\mathbf{O}(1)$ neighborhood of $\lambda_1 \in S_1$ and define

$$\Lambda = \{[B \cap S_1] \setminus (U_1 \cup N)\} \cup \tilde{K}, \qquad \Lambda_H = \{[B_H \cap S_1] \setminus (U_1 \cup N)\} \cup \tilde{K} \tag{3.17}$$

where B is the domain in (1.16) and $B_H = B \cap \{|\lambda| \leq H\}$ with H as in Lemma 3.2. Thus Λ contains the unstable half plane with a small neighborhood U_1 of λ_1 removed, together with a contour on $\mathcal{R}$ enclosing the neighborhood $\tilde{N}$ of $\lambda_*(\delta)$ in its interior. The latter neighborhood includes the translational eigenvalue at $\lambda = 0$ for all small δ.

By the above choice of (3.16), there exists a unique solution $\hat{y}_{1R}(\xi, \tilde{\lambda})$ of the projectivized version of (3.16) such that

$$\lim_{\xi \to -\infty} \hat{y}_{1R}(\xi, \tilde{\lambda}) = \hat{e}_{1R}(\tilde{\lambda})$$

and

$$\lim_{\xi \to +\infty} \hat{y}_{1R}(\xi, \tilde{\lambda}) = \hat{e}_{1R}(\tilde{\lambda}).$$

We remark that the condition at $-\infty$ uniquely determines $\hat{y}_{1R}(\xi, \tilde{\lambda})$, while that at $+\infty$ guarantees $\tilde{\lambda}$ is not an eigenvalue of (3.16). Moreover, this limit is uniform over compact subsets of $\tilde{\lambda}$. Thus, if $K \subset \Lambda$ is a simple closed curve, there exists $T_R \geq T$ independent of $\tilde{\lambda} \in K$ such that $\hat{y}_{1R}(T_R, \tilde{\lambda}) \in V_0(\tilde{\lambda})$. Here, the time T is as in Lemma 3.4.

The goal now, colloquially speaking, is to show that, for the full system, the positively invariant neighborhood over $T_R \leq \xi < \infty$ flares out like an elephant's trumpet as it gets pulled back to some time $T_L < -T$. If this goal can be achieved, then one has constructed a positively invariant set $\Omega \subset \mathbb{CP}^3 \times \mathbb{R}$ relative to $(-\infty, \infty)$. To be precise, we introduce the following notation.

DEFINITION 3.5. Let $\varphi(\hat{y}_0, \xi, \delta)$ denote the flow map for the initial value problem (2.18) with $\hat{y}(T_R) = \hat{y}_0$. Explicitly, $\varphi(\hat{y}_0, \xi, \delta) = \hat{y}(\xi)$. Also, let $\varphi_R(\hat{y}_0, \xi) \equiv \varphi(\hat{y}_0, \xi, 0)$.

The main estimate required in this construction is to determine $T_L \leq -T$ such that $V_0(\tilde{\lambda}) \subset \varphi(V_0(\tilde{\lambda}), T_L, \delta)$ for $\delta < \delta_0$ (see (3.21), below). To this end, we introduce the new independent variable τ defined by

$$\xi(\tau) \equiv \frac{1}{2\kappa} \ln \left(\frac{1+\tau}{1-\tau} \right) \tag{3.18}$$

for $-1 < \tau < 1$ and working on the compactified autonomous version of (3.6):

$$\begin{aligned} \hat{y}' &= \hat{a}_R(\tau, \tilde{\lambda}, \hat{y}) \\ \tau' &= \kappa(1-\tau^2), \end{aligned} \tag{3.19}$$

where $\hat{a}_R(\tau, \tilde{\lambda}, \hat{y}) = \hat{a}_R(\xi(\tau), \tilde{\lambda}, \hat{y})$. The above analysis of (3.6) immediately implies that $(\hat{e}_{1R}(\tilde{\lambda}), -1)$ is a rest point of (3.19) with a one (real) dimensional unstable manifold, and, for $\tau > -1$, the solution on $W^u(\hat{e}_{1R}(\tilde{\lambda}), -1)$ is precisely $\hat{y}_{1R}(\xi, \tilde{\lambda})$. We also need to consider the asymptotic system at $-\infty$:

$$\begin{aligned} \hat{y}' &= \hat{a}_R(-1, \tilde{\lambda}, \hat{y}) \\ \tau &= -1. \end{aligned} \tag{3.20}$$

The projectivized eigenvector $\hat{e}_{1R}(\tilde{\lambda})$ is an attracting rest point of (3.20). Hence, we may consider the attracting neighborhood $V_0(\tilde{\lambda}) \subset \mathbb{CP}^3$ of $\hat{e}_{1R}(\tilde{\lambda})$ of this flow determined as in (3.15). Thus, the ω-limit set of $(V_0(\tilde{\lambda}), -1)$ under (3.20) is $\{(\hat{e}_{1R}(\tilde{\lambda}), -1)\}$.

LEMMA 3.6. *Suppose that $T = T(\frac{\alpha}{2}, H), > 0$ is as in Lemma 3.4. There exists a $T_L \leq -T$ such that $V_0(\tilde{\lambda}) \subset \varphi_R(V_0(\tilde{\lambda}), T_L)$ for all $\tilde{\lambda} \in \Lambda_H$.*

PROOF. A similar lemma is proved in **[12]**, so we only provide a sketch. The key points are that (i) $\hat{y}_{1R}(T_R, \lambda) \in V_0(\tilde{\lambda})$, (ii) the α-limit set $\alpha(\hat{y}_{1R}(T_R, \tilde{\lambda}), \tau_R) = (\hat{e}_{1R}(\tilde{\lambda}), -1)$ under φ_R where $\tau_R = \tau(T_R)$, and this is the only such orbit of (3.19) , and (iii) the α-limit set of $(V_0(\tilde{\lambda}), T_R)$ under φ_R is the entire space $(\mathbb{CP}^3, -1)$.

The last fact is a consequence of the observations that $\hat{e}_{1R}(\tilde{\lambda})$ is an attracting rest point of the asymptotic flow (3.20) and that the α-limit set of $(V_0(\tilde{\lambda}), \tau_R)$ under φ_R includes a full neighborhood of the rest point $(\hat{e}_{1R}(\tilde{\lambda}), -1)$. The latter observation follows from our choice of T_R, which ensures that $\hat{y}_{1R}(T_R, \tilde{\lambda}) \in V_0(\tilde{\lambda})$.

The constant $H > 0$ appears in (3.13) as a bound for $|\lambda|$ and $|z|$ for $z \in V_0(\tilde{\lambda})$. Hence $T(\alpha/2, H), \delta(\alpha/2, H)$ can be chosen uniformly for $\tilde{\lambda}$ in compact subsets of Λ. See [**12**] for additional details. □

This lemma actually implies $V_0(\tilde{\lambda}) \subset \varphi_R(V_0(\tilde{\lambda}), \xi)$ for all $\xi \leq T_L$, since by Lemma 3.4, $V_0(\tilde{\lambda}) \times [-1, \tau(T_L)]$ is positively invariant relative to $[-1, \tau(T_L)]$. Hence, in the reduced system, the pullback of the invariant set relative to $[T, \infty]$ flares out over $\mathbb{CP}^3$ and therefore the desired positively invariant set has been constructed.

We now show the full system also has a positively invariant set. Let $V_0(\tilde{\lambda})$ be as above, set $\alpha > 0$ as in (3.15), and δ_0 and T as in Lemma 3.4. A tube $\Omega_1 \subset \mathbb{CP}^3 \times \mathbb{R}$ is now defined by

(3.21)

$$\Omega_1 = \left(V_0(\tilde{\lambda}) \times (-\infty, T_L]\right) \cup \left(\bigcup_{T_L \leq \xi \leq T_R} (\varphi(V_0(\tilde{\lambda}), \xi, \delta), \xi) \right) \cup \left(V_0(\tilde{\lambda}) \times [T_R, \infty)\right).$$

THEOREM 3.7. *There exists a $\delta_0 > 0$, such that for all $0 < \delta < \delta_0$ and $\tilde{\lambda} \in \Lambda_H$, the tube Ω_1 is positively invariant for the flow consisting of (2.18) augmented with $\xi' = 1$, relative to the full space $\mathbb{CP}^3 \times \mathbb{R}$.*

PROOF. From Lemma 3.4, we know that $\delta < \delta_0$ implies that $V_0(\tilde{\lambda}) \times (-\infty, T_L]$ is positively invariant under φ relative to $(-\infty, T_L]$ and that $V_0(\tilde{\lambda}) \times [T_R, +\infty)$ is positively invariant relative to $[T_R, +\infty)$. Since $\varphi(V_0(\tilde{\lambda}), T_R, \delta) = V_0(\tilde{\lambda})$, the theorem follows if it can be shown that $\varphi(V_0(\tilde{\lambda}), T_L, \delta) \subset V_0(\tilde{\lambda})$ for all $\delta < \delta_0$.

First, we set $\alpha > 0$ as in (3.15). This condition only depends on the asymptotic matrix $a_R(\tilde{\lambda})$. In the limit $\delta \to 0$, the perturbed coefficients $u_o(\xi, \delta)$ and $v_o(\xi, \delta)$ approach their singular limits $\bar{c}$ and $S(\xi)$ given by (1.7) uniformly on compact ξ–intervals. More precisely, Lemma 3.1 provides the estimate (3.13) for $|F_R - F|$ in local coordinates. In the latter estimate, H depends only on $\lambda \in K$. We then set $r = \alpha/2$ in (3.13) as in Lemma 3.4 to determine T and δ_0. This now determines $T_L < 0 < T_R$ as above relative α. With T_L and T_R now fixed uniformly for all $\delta < \delta_0$, the theorem now follows from the continuity of the flow φ in δ on compact intervals, together with result of Lemma 3.6 for the $\delta = 0$ flow. This last step may require resetting δ_0 if necessary to achieve the desired result. □

3.4. Corollaries of the Fast Tracking Theorem

The results of the preceding section can now be used to construct various bundles.

COROLLARY 3.8. *(i) Let $\hat{y}_1(\xi, \tilde{\lambda}, \delta)$ be the unique solution of (2.18) tending to $\hat{e}_1(\tilde{\lambda})$ as $\xi \to -\infty$. There exists $\delta_0(\Lambda) > 0$ such that*

$$\lim_{\xi \to +\infty} \hat{y}_1(\xi, \tilde{\lambda}, \delta) = \hat{e}_1(\tilde{\lambda}) \tag{3.22}$$

uniformly for $\delta < \delta_0(\Lambda)$ and for all $\tilde{\lambda} \in \Lambda$, where Λ is as in (3.17).

(ii) Let $y_1(\xi,\tilde{\lambda},\delta) \in \hat{y}_1(\xi,\tilde{\lambda},\delta)$ be the (unique) solution of (2.4) satisfying the normalization condition in Lemma 2.1. Then there exists an analytic function $R_1(\tilde{\lambda},\delta)$ on Λ and $m_1 > Re\,\mu_1(\tilde{\lambda})/2 > 0$ such that

$$y_1(\xi,\tilde{\lambda},\delta) = e^{\mu_1\xi}[R_1(\tilde{\lambda},\delta)e_1(\tilde{\lambda}) + \mathbf{O}(e^{-m_1\xi})]$$

for $\xi \geq 0$, uniformly for $\tilde{\lambda} \in \Lambda$ and $\delta < \delta_0(\Lambda)$.

PROOF. The first assertion follows immediately from Lemma 3.2, Theorem 3.7, and standard theorems about ω-limit sets. Since μ_1 has real part uniformly larger than the real parts of all the other eigenvalues, (ii) then follows from Lemma 2.5 in [**14**]. □

Now let $K \subset \Lambda$ be a simple closed curve, so that the set of eigenvalues of (3.16) in Λ, namely $\{0, \lambda_1\}$, together with the branch point $\lambda_*(\delta)$ are a positive distance from K for all sufficiently small $\delta > 0$. We consider two distinct situations: (i) $K \subset \Lambda \setminus \tilde{N}$, or (ii) $K = \tilde{K} = \partial\tilde{N}$. In the latter case, note that $\tilde{K}$ encloses the branch point $\lambda_*(\delta)$ and the origin $\lambda = 0 \in S_1$ in its interior for all small $\delta > 0$.

In either case, it follows from Corollary 3.8 that the clutch construction described in Chapter 2, section 3 can now be used to define a 1-dimensional bundle $\mathcal{E}_f(K,\delta)$. In particular, we set $\alpha(\tau,\tilde{\lambda},\delta) = e^{-\mu_1\xi}y_1(\xi(\tau),\tilde{\lambda},\delta)$, $\beta(\tilde{\lambda},\delta) = e_1(\tilde{\lambda})$, and $\gamma(\tilde{\lambda},\delta) = \mathbf{O}(e^{-m_1\xi})$ lying in the space spanned by $e_i, i = 2,3,4$. The clutching function for $\mathcal{E}_f(K,\delta)$ is the transmission coefficient $R_1(\tilde{\lambda},\delta)$ of the previous corollary.

Since $K \subset \Lambda$ we observe that the projectivized solution $(\hat{y}_{1R}(\xi), \tau(\xi))$ of (3.19) is an orbit connecting $(\hat{e}_{1R}(\tilde{\lambda}), -1)$ to $(\hat{e}_{1R}(\tilde{\lambda}), +1)$ for $\tilde{\lambda} \in K$. Select the unique solution $y_{1R}(\xi,\tilde{\lambda}) \in \hat{y}_{1R}(\xi,\tilde{\lambda})$ of (3.16) satisfying the normalization $e^{-\mu_1\xi}y_{1R}(\xi,\tilde{\lambda}) = e_{1R}(\tilde{\lambda})$ as $\xi \to -\infty$. By Lemma 2.5 of [**14**] we have that

$$y_{1R}(\xi,\tilde{\lambda}) = e^{\mu_1\xi}[R_{1R}(\tilde{\lambda})e_{1R} + \mathbf{O}(e^{-m_1\xi})]$$

as $\xi \to +\infty$, where $m_1 > 0$, uniformly for $\tilde{\lambda} \in K$. Hence we can also construct a reduced bundle $\mathcal{E}_{fisher}(K)$ for the linearized equations associated to the reduced Fisher wave uniformly for $\tilde{\lambda} \in K$.

COROLLARY 3.9. *(i) Let $\xi_+(\delta) = 2/\delta^s$ with $s > 0$, let $K \subset \Lambda \setminus \tilde{N}$, and let $\mathcal{E}_f(K,\delta)$ be the bundle constructed above. There exists $\delta_0(\Lambda) > 0$ such that for all $\delta < \delta_0(\Lambda)$,*

$$c_1(\mathcal{E}_f(K,\delta)) = \begin{cases} 0 & \text{if } \lambda_1 \text{ is exterior to } K \\ 1 & \text{if } \lambda_1 \text{ is exterior to } K. \end{cases}$$

(ii) Let $K = \tilde{K} = \partial\tilde{N}$; then $c_1(\mathcal{E}_f(\tilde{K},\delta)) = 2$ for all $\delta < \delta_0(\Lambda)$.

PROOF. By Theorem 3.7, the bundles $\mathcal{E}_f(K,\delta), \mathcal{E}_f(\tilde{K},\delta)$ persist for all sufficiently small $\delta > 0$. Their Chern numbers are therefore independent of δ. Their topology is determined by the clutching function $R_1(\tilde{\lambda},\delta)$ so that it suffices to show that $R_1(\tilde{\lambda},\delta) \to R_{1R}(\tilde{\lambda})$ as $\delta \to 0$, uniformly along each contour.

We first show that

$$\lim_{\delta\to 0} \hat{y}_1(\xi,\tilde{\lambda},\delta) = \hat{y}_{1R}(\xi,\tilde{\lambda}) \tag{3.23}$$

holds *uniformly* for all $\xi \in \mathbb{R}$ and $\tilde{\lambda} \in \Lambda$. Hence given $\epsilon > 0$ we determine $\delta_o(\epsilon)$ such that $\rho[\hat{y}_1(\xi,\tilde{\lambda},\delta), \hat{y}_{1R}(\xi,\tilde{\lambda})] < \epsilon$ for all $\delta < \delta_o(\epsilon)$.

For $|\lambda| \geq H$, it follows from Lemma 3.2 that $hy_1(\xi, \lambda, \delta)$ remains in a small neighborhood of $\hat{e}_1(\lambda)$ for all ξ, where the size of the neighborhood can be made arbitrarily small by choosing H large and δ_o sufficiently small. Hence for given ϵ, there exists $H(\epsilon) \geq H$ and $d_o(\epsilon) < d_o$ such that $\rho[\hat{y}_1(\xi, \lambda, \delta), \hat{y}_{1R}(\xi, \lambda)] < \epsilon$ for all ξ and for $|\lambda| \geq \epsilon$ and $\delta < \delta_o$.

With $H(\epsilon)$ now fixed, we now need only derive the estimate $\rho[\hat{y}_1, \hat{y}_{1R}] < \epsilon$ for $\tilde{\lambda}$ on the compact set $\Lambda \setminus \{|\lambda| \geq H(\varepsilon)\}$. The positively invariant neighborhood Ω_1 of the reduced orbit $(\hat{y}_{1R}(\xi, \tilde{\lambda}), \xi)$ is determined by first setting the size of the (small) neighborhood $V_0(\tilde{\lambda})$ of $\hat{e}_{1R}(\tilde{\lambda})$. The large but $\mathbf{O}(1)$ times $T_L < 0 < T_R$ defining the transitional region in (3.21) are then determined relative to the size of $V_0(\tilde{\lambda})$; i.e. $T_R, -T_L \to +\infty$ as the diameter of $V_0(\tilde{\lambda})$ tends to zero.

If (3.23) were false, there would exist times ξ_n such that

$$\rho[\hat{y}_1(\xi_n, \tilde{\lambda}_n, \delta_n), \hat{y}_{1R}(\tilde{\lambda}_n, \delta_n)] \geq A$$

for some $A > 0$ and some sequences $\tilde{\lambda}_n \in K, \tilde{K}$ and $\delta_n \to 0$. Passing to a subsequence, it can be assumed that $\tilde{\lambda}_n$ converges to a limit $\tilde{\lambda}$. By choosing A smaller if necessary, it can be assumed that $\rho[\hat{e}_{1R}(\tilde{\lambda}), \hat{e}_{jR}(\tilde{\lambda})] \geq A/2$ for $j = 2, 3, 4$. Set the neighborhood $V_0(\tilde{\lambda})$ so small that $\rho[V_0(\tilde{\lambda}), \hat{e}_{1R}(\tilde{\lambda})] < A/2$. This neighborhood determines an interval $[T_L, T_R]$ independent of n, as in the proof of Theorem 3.7. By Theorem 3.7, $(\hat{y}_1(\xi_n, \tilde{\lambda}_n, \delta_n), \xi_n) \in \Omega_1$ for all sufficiently large n. We therefore have that $\xi_n \in [T_L, T_R]$ for all large n. By passing to a subsequence if necessary, it can then be assumed that ξ_n converges to a limit $\xi_0 \in [T_L, T_R]$, and that $\hat{y}_1(\xi_n, \tilde{\lambda}_n, \delta_n)$ converges to a limit $\hat{y}_*$ with $\rho[\hat{y}_*, \hat{y}_{1R}(\xi_0)] \geq A$. Since $(\hat{y}_{1R}(\xi, \tilde{\lambda}), \tau(\xi))$ is the unique solution of (3.19) which tends to $(\hat{y}_{1R}(\tilde{\lambda}), -1)$ as $\xi \to -\infty$ and $\hat{y}_* \neq \hat{y}_{1R}(\xi_0, \tilde{\lambda})$, it follows that the solution $(\hat{y}_{\mathbb{R}}(\xi, \tilde{\lambda}), \tau(\xi))$ of (3.19) leaves an $A/2$ neighborhood of $(\hat{e}_{1R}(\tilde{\lambda}), -1)$ for sufficiently large and negative $\xi_1 < \xi_0$. It therefore follows from standard continuous dependence theorems that $\hat{y}_1(\xi_1, \tilde{\lambda}_n, \delta_n)$ also lies at a distance of at least $A/2$ from $\hat{e}_{1R}(\tilde{\lambda})$ for sufficiently large n, contradicting the fact that $(\hat{y}_1(\xi, \tilde{\lambda}, \delta), \xi) \in \Omega_1$ for all ξ. This establishes (3.23).

Next consider $y_1 \in \hat{y}_1$ and $y_{1R} \in \hat{y}_{1R}$ of the original linear equations which satisfy the asymptotic conditions

$$\begin{aligned} y_1(\xi, \tilde{\lambda}, \delta) &= e^{\mu_1 \xi}[e_1(\tilde{\lambda}) + \mathbf{O}(e^{-m_1 \xi})] \\ y_{1R}(\xi, \tilde{\lambda}) &= e^{\mu_1 \xi}[e_{1R}(\tilde{\lambda}) + \mathbf{O}(e^{-m_1 \xi})] \end{aligned}$$

as $\xi \to -\infty$. It follows from Corollary 3.8 above and Lemma 2.5 in **[14]** that there exist uniquely determined coefficients $\tilde{R}_1(\tilde{\lambda}, \delta)$ and $\tilde{R}_{1R}(\tilde{\lambda})$ which are analytic on Λ and such that

$$y_1(\xi, \tilde{\lambda}, \delta) = e^{\mu_1 \xi}[\tilde{R}_1(\tilde{\lambda}, \delta) e_1(\tilde{\lambda}) + \mathbf{O}(e^{-m_1 \xi})] \tag{3.24}$$

$$y_{1R}(\xi, \tilde{\lambda}) = e^{\mu_1 \xi}[\tilde{R}_{1R}(\tilde{\lambda}) e_{1R}(\tilde{\lambda}) + \mathbf{O}(e^{-m_1 \xi})] \tag{3.25}$$

uniformly for *all* $\xi \geq 0$. Since $\hat{y}_1$ converges uniformly to $\hat{y}_{1R}$ for all $\xi \in \mathbb{R}$ and $\tilde{\lambda} \in K$, it therefore follows that $\tilde{R}_1(\tilde{\lambda}, \delta)$ converges uniformly to $\tilde{R}_{1R}(\tilde{\lambda})$ as $\delta \to 0$.

Hence, it suffices to compute $c_1(\mathcal{E}_{fisher}(K))$ and $c_1(\mathcal{E}_{fisher}(\tilde{K}))$. It then follows from an argument similar to (but simpler than) that of Proposition 2.8 that the latter two quantities are $W(D_{fisher}(K))$ and $W(D_{fisher}(\tilde{K}))$ respectively, where $D_{fisher}(\tilde{\lambda})$ is the Evans function for the Fisher wave, since $\xi_+(\delta) = 2/\delta^s$. Thus, the Chern numbers of these bundles are the winding numbers of $D_{fisher}(K)$ and

$D_{fisher}(\tilde{K})$, respectively. It then follows from **[1]** in case (i) of the Corollary that $c_1(\mathcal{E}_{fisher}(K))$ is the multiplicity of eigenvalues of the Fisher wave inside K, which is either zero or one. In case (ii), the function $R_{1R}(\tilde{\lambda})$ is analytic in $\tilde{N}$ and has the *same* value on each sheet of the Riemann surface, since $\lambda_*(\delta)$ is not a branch point for $e_{1R}(\tilde{\lambda})$. Since $R_{1R}(0) = 0$ where $0 \in S_1$, it follows that $R_{1R}(\tilde{0}) = 0$ as well, where $\tilde{0} \in S_2$ is the point on S_2 over $\lambda = 0 \in S_1$. This accounts for all the roots of $R_{1R}(\tilde{\lambda})$ in $\tilde{N}$, so that $c_1(\mathcal{E}_{fisher}(\tilde{K})) = 2$.

Evaluating (3.24) at $\xi = 2/\delta^s$, it then follows from (3.23) that $R_1(\tilde{\lambda}, \delta)$ converges uniformly to $R_{1R}(\tilde{\lambda})$ along each contour, so that both statements in the Corollary follow from the calculations for the Fisher bundles. □

REMARK 3.10. It should be emphasized that these results in themselves are insufficient to draw any conclusions regarding the eigenvalue count for the perturbed wave for $\delta > 0$ inside K or $\tilde{K}$, since the slow summand of $\mathcal{E}(K, \delta)$ may have a negative Chern number.

We conclude with one further observation regarding constructions analogous to those of Lemma 3.2 and Theorem 3.7 for the solution $\hat{y}_4$.

COROLLARY 3.11. *(i) Let $W_o(\omega) = V_o(\omega) + (0, 0, -2\omega)$, where $V_o(\omega)$ is as in Lemma 3.2. Then the neighborhood $W_o(\omega)$ is negatively invariant for (3.11) for $\delta < \delta_o$ and $d < d_o$.*

(ii) For $\delta < \delta_0$ and $\tilde{\lambda} \in \Lambda \cap \{|\lambda| \leq H\}$, there exists a negatively invariant tube Ω_4 which is constructed as in (3.21) with $\hat{e}_1$ replaced by $\hat{e}_4$, and where the flow map φ in Definition 3.2 is has its initial condition at $\xi = T_L$.

(iii) There exists $\delta_0(\Lambda) > 0$ such that the solution $\hat{y}_4$ which is (uniquely) determined by the condition $\hat{y}_4 \to \hat{e}_4$ as $\xi \to +\infty$ satisfies

$$\lim_{\xi \to -\infty} \hat{y}_4(\xi, \tilde{\lambda}, \delta) = \hat{e}_4(\tilde{\lambda})$$

uniformly for $\delta < \delta_0(\Lambda)$ and for $\tilde{\lambda} \in \Lambda$.

The proofs are the same as those of Lemma 3.2 and Theorem 3.7 after a time reversal.

CHAPTER 4

The Slow Subbundle

In this chapter we define and estimate a second solution of the projectivized equations (2.18). The goal is to find a solution $\hat{y}_2(\xi, \tilde{\lambda}, \delta)$ that is asymptotic to $\hat{e}_2(\tilde{\lambda}, \delta)$ as $\xi \to -\infty$, and which remains near the *slow subspace*,

$$\Sigma = \text{span}\{e_2(\tilde{\lambda}, \delta), e_3(\tilde{\lambda}, \delta)\} \tag{4.1}$$

outside some neighborhood of the transition layer, $|\xi| \geq 2/\delta^s$, where $0 < s < 1$. (The factor "2" is included for technical reasons.) Note that Σ is the (u, p) plane for all $(\tilde{\lambda}, \delta)$. The estimates are obtained along certain simple, closed contours $K \subset \Lambda$, where Λ is as in (3.17); (see Lemmas 4.3, 4.4 below).

4.1. Definition of the Slow Solution

First, however, we need to define a suitable candidate for this solution, since most solutions which are asymptotic to $\hat{e}_2$ in the backwards direction will not have this property. To this end, we introduce a 3-dimensional subspace of solutions $\Theta(\xi, \tilde{\lambda}, \delta)$ whose defining condition is at $+\infty$.

LEMMA 4.1. *Suppose that $\delta < \delta_o$, where δ_o is as in Theorem 3.7 and that $\tilde{\lambda} \in \Lambda$. There is then a uniquely determined 3-dimensional subspace of solutions $\Theta(\xi, \tilde{\lambda}, \delta)$ of (2.2) such that*

$$\lim_{\xi \to +\infty} \Theta(\xi, \tilde{\lambda}, \delta) = \textit{span}\{e_2, e_3, e_4\}(\tilde{\lambda}, \delta).$$

PROOF. There exists some $c > 0$ such that

$$\text{Re}\, \mu_1(\tilde{\lambda}) > c > \max\{\text{Re}\, \mu_i(\tilde{\lambda}, \delta) : i = 2, 3, 4\}$$

for all δ and $\tilde{\lambda}$ as in the statement of the lemma. The existence of the subspace $\Psi(\xi, \tilde{\lambda}, \delta)$ then follows from Lemma 3.3 in [**1**]. □

LEMMA 4.2. *For sufficiently small $\delta > 0$ and $\tilde{\lambda} \in \Lambda$, there exists a nontrivial solution $y_2(\xi, \tilde{\lambda}, \delta)$ of (2.4) such that*

$$\begin{aligned} \lim_{\xi \to -\infty} \hat{y}_2(\xi, \tilde{\lambda}, \delta) &= \hat{e}_2(\tilde{\lambda}, \delta) \\ \lim_{\xi \to +\infty} \rho[\hat{y}_2(\xi, \tilde{\lambda}, \delta), \hat{\Sigma}] &= 0, \end{aligned} \tag{4.2}$$

where $\rho[\ ,\ ]$ is the metric on $\mathbb{CP}^3$ introduced below (2.18). The solution $\hat{y}_2(\xi, \tilde{\lambda}, \delta)$ of (2.18) satisfying (4.2) is unique.

PROOF. Let $\Phi(\xi, \tilde{\lambda}, \delta)$ be as in Lemma 2.5. We claim that $\Phi(\xi, \tilde{\lambda}, \delta)$ and $\Theta(\xi, \tilde{\lambda}, \delta)$ intersect nontrivially and transversely in $\mathbb{C}^4$. The nontriviality follows from a simple dimension count. Since the solution $\hat{y}_1(\xi, \tilde{\lambda}, \delta)$ of (2.18) is uniquely determined by its defining condition at $-\infty$, it follows that $\hat{y}_1(\xi, \tilde{\lambda}, \delta) \in \hat{\Phi}(\xi, \tilde{\lambda}, \delta)$ for all ξ. However, by Corollary 3.8, it follows that $\hat{y}_1(\xi, \tilde{\lambda}, \delta)$ is near $\hat{e}_1(\tilde{\lambda})$ in $\mathbb{CP}^3$ for ξ near $+\infty$. Since $e_1(\tilde{\lambda})$ is transverse to the subspace $\text{span}\{e_i(\tilde{\lambda}, \delta) : i = 2, 3, 4\}$, it follows that $\dim\left(\Phi(\xi, \tilde{\lambda}, \delta) \cap \Theta(\xi, \tilde{\lambda}, \delta)\right) \leq 1$.

We therefore see that there exists a unique solution $\hat{y}_2(\xi, \tilde{\lambda}, \delta)$ of (2.18) such that $\hat{y}_2(\xi, \tilde{\lambda}, \delta) \in \hat{\Phi}(\xi, \tilde{\lambda}, \delta) \cap \hat{\Theta}(\xi, \tilde{\lambda}, \delta)$. All solutions in $\hat{\Phi}(\xi, \tilde{\lambda}, \delta)$ are backwards asymptotic to either $\hat{e}_1$ or $\hat{e}_2$. Since $\hat{y}_1$ is backwards asymptotic to $\hat{e}_1$ and $\hat{y}_2 \neq \hat{y}_1$, the first condition in (4.2) follows from the uniqueness of $\hat{y}_1(\xi, \tilde{\lambda}, \delta)$.

The behavior of $\hat{y}_2(\xi, \tilde{\lambda}, \delta)$ is also determined to some extent at $+\infty$. The simplest situation is when $\tilde{\lambda}$ is not on the branch cut of $\mathcal{R}$. In this case, the matrix $a_0(\tilde{\lambda}, \delta)$ is hyperbolic with four eigenvalues with distinct, nonzero real parts, when $\delta > 0$. The system $\hat{y}' = \hat{a}_0(\tilde{\lambda}, \delta, \hat{y})$ is therefore a Morse-Smale system in which all solutions are either rest points or connections $\hat{e}_j \to \hat{e}_i$. If the full equations (2.18) are augmented by τ as in (3.19), then it follows from standard theorems about ω-limit sets that $\lim_{\xi \to +\infty} \hat{y}_2(\xi, \tilde{\lambda}, \delta)$ must be one of the rest points, $\hat{e}_i(\tilde{\lambda}, \delta)$.

We now show that $\lim_{\xi \to +\infty} \hat{y}_2$ is either $\hat{e}_2$ or $\hat{e}_3$. By the previous paragraph, the only other possibilities are that this limit is either $\hat{e}_1$ or $\hat{e}_4$. The limit cannot be $\hat{e}_1$, since that would violate the defining condition for solutions in $\hat{\Theta}$ at $+\infty$.

The only other possibility is that $\lim_{\xi \to +\infty} \hat{y}_2 = \hat{e}_4$. However, there is a unique solution of (2.18) with this limit, and by Corollary 3.11, the limit as $\xi \to -\infty$ of this solution must also be $\hat{e}_4$. This violates the defining condition for $\hat{\Phi}$ at $-\infty$.

In the case that $\tilde{\lambda}$ lies along the branch cut, the matrix $a_0(\tilde{\lambda}, \delta)$ is no longer hyperbolic. It has a pair of imaginary eigenvalues which are complex conjugates. In this case, the asymptotic flow is still a Morse-Smale flow, however $\hat{\Sigma}$ is now a nontrivial, hyperbolic periodic orbit or a circle of rest points (at $\lambda = \lambda_*(\delta)$). This is because the real parts of $\mu_{1,4}(\tilde{\lambda})$ are bounded away from zero, while $\mu_{2,3}(\tilde{\lambda}, \delta)$ are both $\mathbf{O}(\delta)$ and have zero real parts. The same argument as in the previous case shows that the ω-limit set of $\hat{y}_2$ must still lie in $\hat{\Sigma}$. □

4.2. Tracking the Slow Solution over the Slow Fields

Next, we characterize the behavior of $\hat{y}_2(\xi, \tilde{\lambda}, \delta)$ in the outer layers. In contrast to the fast tracking in Chapter 3, we now define the left and right slow fields to be $\xi \leq -2/\delta^s$ and $2/\delta^s \leq \xi$, respectively, where $s \in (0, 1-\beta)$ is fixed as $\delta \to 0$. In Lemma 4.9 we ultimately choose $s = s(\beta) = (1-\beta)/3$. Exponential estimates are obtained later in this chapter over the fast field. This is crucial, since the matching conditions for the NLEP equation equate terms that are algebraically small in δ. However, it is difficult to obtain such estimates that are uniform on Λ, since this is an unbounded domain. We first prove an estimate for large $\lambda \in \Lambda$ that allows us to restrict to the case where λ is at worst algebraically large.

LEMMA 4.3. *Suppose that $s \in (0, 1-\beta)$, that $h > 0$ is a given $\mathbf{O}(1)$ constant, and that $\lambda \in \Lambda$ with $|\lambda|^{1/2} \geq h\delta^{s-1}$. Then there exists $\delta_o > 0$ such that λ is* not *an eigenvalue of the perturbed wave for all $\delta < \delta_o$.*

PROOF. We work directly with the eigenvalue problem (1.9) for the perturbed operator L. The equations for L are

$$L\begin{pmatrix} u \\ v \end{pmatrix} = \begin{pmatrix} b\delta^2\lambda u \\ b\lambda v \end{pmatrix}.$$

We have that $\delta^2|\lambda| \geq \delta^{2s}h^2$. Suppose that (u,v) is an eigenfunction of L for $\lambda \in \Lambda$ with $|\lambda| \geq h^2\delta^{2(s-1)}$. Thus $|u|, |v| \to 0$ exponentially as $|\xi| \to \pm\infty$. Let $\lambda = \lambda_R + i\lambda_I$, $u = u_R + iu_I$, and $v = v_R + iv_I$, multiply (1.9) by $(\bar{u}, \bar{v})$, and integrate to obtain

$$\begin{aligned} -\int_{-\infty}^{\infty} [|u'|^2 + |v'|^2]\, d\xi &= \int_{-\infty}^{\infty} \{[b\delta^2\lambda_R + \delta^{2-2\beta}v_0^2 + \delta^{4-\beta}a]|u|^2 \\ &\quad + [b\lambda_R + b - 2u_o v_o]|v|^2 \\ &\quad + [2\delta^{2-2\beta}u_o v_o - v_o^2](u_R v_R + u_I v_I)\}\, d\xi \\ 0 &= \int_{-\infty}^{\infty} \{[v_o^2 + 2u_o v_o \delta^{2-2\beta}](v_I u_R - u_I v_R) \\ &\quad + b\delta^2\lambda_I|u|^2 + b\lambda_I|v|^2\}\, d\xi. \end{aligned} \tag{4.3}$$

Since $\lambda \in \Lambda$ and $|\lambda| \geq h^2\delta^{2(s-1)}$, it follows that $\max\{\lambda_R, |\lambda_I|\} > h^2\delta^{2(s-1)}/\sqrt{2}$ for sufficiently small δ. Here, we have used that the argument θ of $b(1+\lambda)$ satisfies $|\theta| < \pi/2 + \mathbf{O}(d^2)$ for $\lambda \in B$, and that the argument ψ of λ satisfies $|\psi - \phi| = \mathbf{O}(\delta^{2(1-s)})$.

Assume first that this maximum occurs for λ_R, so that $b\delta^2\lambda_R \geq bh^2\delta^{2s}/\sqrt{2}$. Since $s < 1-\beta$, it follows that $b\delta^2\lambda_R + \delta^{2-2\beta}v_0^2 + \delta^{4-\beta}a \geq h^2 b\delta^{2s}/2$ and that $b\lambda_R + b - 2u_o v_o \geq bh^2\delta^{2(s-1)}/2$ for all sufficiently small $\delta > 0$. Next, we have that for any $\epsilon > 0$,

$$[2\delta^{2-2\beta}u_o v_o - v_o^2](u_R v_R + u_I v_I) \leq K[\epsilon|u|^2 + \frac{1}{\epsilon}|v|^2],$$

where $K > 0$ is an upper bound for the coefficient in brackets. We then obtain

$$-\int_{-\infty}^{\infty} [|u'|^2 + |v'|^2]\, d\xi \geq (bh^2/2)\int_{-\infty}^{\infty} \left\{[\delta^{2s} - K\epsilon]|u|^2 + [\delta^{2(s-1)} - \frac{K}{\epsilon}]|v|^2\right\} d\xi.$$

In order to make both coefficients of the integral on the right side positive, we must choose $\epsilon < \delta^{2s}/K$ and $\epsilon > K\delta^{2(1-s)}$, i.e. $K\delta^{2(1-s)} < \epsilon < \delta^{2s}/K$. Such a choice is always possible for sufficiently small δ provided that $s < 1-s$, i.e. $s < 1/2$. Since $s = (1-\beta)/3 < 1/3$, there exists $\epsilon > 0$ for all $\beta \in [0,1)$ such that this inequality holds. It therefore follows that u, v must vanish identically.

In the second case where $|\lambda_I| \geq h^2\delta^{-2r-2\beta}/\sqrt{2}$, we use the second equation in (4.3) together with a similar estimate to see that again $u \equiv v \equiv 0$. □

The search for eigenvalues can now be restricted to the domain

$$\Lambda_h = \Lambda \cap \{\lambda \in S_1 : |\lambda|^{1/2} \leq h\delta^{s-1}\}. \tag{4.4}$$

Let $\tilde{\lambda} = (\lambda, \pm\sqrt{\lambda - \lambda_*(\delta)})$, and let

$$R = \min\{\min\{\mathrm{Re}\,\sqrt{b(1+\lambda)} : \tilde{\lambda} \in \Lambda\}, \sqrt{b}/2\}. \tag{4.5}$$

Then R is a positive $\mathbf{O}(1)$ constant satisfying $R \leq \sqrt{b}/2$ and $|\mathrm{Re}\,\mu_{1,4}(\tilde{\lambda})| > R$ for all $\tilde{\lambda} \in \Lambda$, with Λ as in (3.17). In the following, it is useful to recall the notational conventions below (2.18).

LEMMA 4.4. *Let $0 < \sigma < s < 1$. There exists $\delta_o > 0$ depending only on σ and s such that*

$$\rho[\hat{y}_2(\xi, \tilde{\lambda}, \delta), \hat{\Sigma}] < e^{-R/\delta^\sigma} \quad (|\xi| \geq 2/\delta^s). \tag{4.6}$$

for all $\delta < \delta_o$ and $\tilde{\lambda} \in \Lambda_h$.

PROOF. In the following, C will denote a positive constant which can be chosen independently of δ and $\tilde{\lambda} \in \Lambda_h$.

The first estimate is that

$$||a(\xi, \tilde{\lambda}, \delta) - a_0(\tilde{\lambda}, \delta)|| < Ce^{-\sqrt{b}/\delta^s}, \tag{4.7}$$

for $|\xi| \geq 1/\delta^s$. Noting that the difference of these two matrices only involves the terms of order $u_o(\xi, \delta)v_o(\xi, \delta)$ and $v_o(\xi, \delta)^2$ and that the difference is independent of λ, the estimate follows from (3.2) in Lemma 3.1.

Our proof is by contradiction. Thus suppose that for some $|\xi_0| \geq 2/\delta^s$ we have that $\hat{y}_2(\xi_0, \tilde{\lambda}, \delta) = \hat{y}_*$, where $\rho[\hat{y}_*, \hat{\Sigma}] = e^{-R/\delta^\sigma}$. Select $y_* \in \hat{y}_*$ with $||y_*|| = 1$. There exist α_i such that $y_* = \sum_{i=1}^4 \alpha_i e_i(\tilde{\lambda}, \delta)$. Since Σ is the (u,p) plane, our normalization of y_* implies that there exists a constant $G > 0$ depending only on the metric such that $\max\{|v|, |q|\} \geq Ge^{-R/\delta^\sigma}$. It then follows that

$$\max\{|\alpha_1 + \alpha_4|, |\sqrt{b(1+\lambda)}||\alpha_1 - \alpha_4|\} \geq Ge^{-R/\delta^\sigma}.$$

However, $|\lambda| \leq h^2\delta^{2(s-1)}$ is at worst algebraically large. It therefore follows that by decreasing R slightly to control the algebraically small factor of $|b(1+\lambda)|^{-1/2}$ coefficient, we have the estimate $\max\{|\alpha_1|, |\alpha_4|\} \geq Ge^{-R/\delta^\sigma}$.

Next, let $d(\tilde{\lambda}) = [\text{Re } (1+\lambda)]^{-1/2}$ where λ is the projection of $\tilde{\lambda}$ on S_1, and let

$$\xi_\pm = \xi_\pm(\tilde{\lambda}) = \xi_0 \pm d(\tilde{\lambda})/\delta^\sigma.$$

It will be shown that there exists $\delta_0(\Lambda_h) > 0$ such that

$$\begin{aligned} \rho[\hat{y}_2(\xi_+, \tilde{\lambda}, \delta), \hat{e}_1(\tilde{\lambda})] &< e^{-R/\delta^\sigma} \text{ if } |\alpha_1| \geq Ge^{-R/\delta^\sigma} \\ \rho[\hat{y}_2(\xi_-, \tilde{\lambda}, \delta), \hat{e}_4(\tilde{\lambda})] &< e^{-R/\delta^\sigma} \text{ if } |\alpha_4| \geq Ge^{-R/\delta^\sigma}, \end{aligned} \tag{4.8}$$

for all $\delta < \delta_0(\Lambda_h)$ and for all $\tilde{\lambda} \in \Lambda_h$.

Suppose that y_* is chosen as above and that $\xi_0 < -2/\delta^s$. It follows from the above that at least one of $|\alpha_1|$ or $|\alpha_4|$ is larger than Ge^{-R/δ^σ}. For definiteness, suppose that this is the case for α_1; the argument for α_4 is the same after a time reversal. In order to establish the validity of (4.8), we compare $\hat{y}_2(\xi, \tilde{\lambda}, \delta)$ with the solution $\hat{y}_0(\xi)$ of the autonomous system

$$\hat{y}_0' = \hat{a}_0(\tilde{\lambda}, \delta, \hat{y}_0), \quad \hat{y}_0(\xi_0) = \hat{y}_*.$$

Let $y_0(\xi)$ be the solution of the associated linear system,

$$y_0' = a_0(\tilde{\lambda}, \delta)y_0, \quad y_0(\xi_0) = y_*,$$

so that

$$y_0(\xi) = \sum_{i=1}^4 \alpha_i e^{\mu_i(\xi-\xi_0)} e_i(\tilde{\lambda}, \delta).$$

Recall that $\alpha_1 = \tilde{G}e^{-R/\delta^\sigma}$ with $|\tilde{G}| \geq G > 0$, where G depends only on the metric ρ. By (4.5) and our choice of R (see (4.5)) and $d = d(\tilde{\lambda})$, we also have that $\text{Re } (-R + \mu_1 d) > R + r$ for some $r > 0$ independent of $\tilde{\lambda} \in \Lambda_h$, and since $\sigma < s$ we also have that $\xi_+ > -1/\delta^s$. These estimates hold for all $\tilde{\lambda} \in \Lambda_h$ and for all small δ. It then follows that we have that

$$y_0(\xi_+) = \tilde{G}e^{(r+R)/\delta^\sigma} e_1(\tilde{\lambda}) + \mathbf{O}(1) \tag{4.9}$$

uniformly for $\tilde{\lambda} \in \Lambda_h$, since $\mu_{2,3}(\tilde{\lambda}, \delta)d(\tilde{\lambda})/\delta^\sigma$ are uniformly of order $\delta^{1-\sigma}$ for $\tilde{\lambda} \in \Lambda_h$ and $\mu_4(\tilde{\lambda})$ has strictly negative real part. Since $|\tilde{G}| \geq G > 0$ where G depends only on ρ and $r > 0$, there exists $\delta_0 > 0$ such that

$$\rho[\hat{y}_0(\xi_+), \hat{e}_1(\tilde{\lambda})] < e^{-R/\delta^\sigma} \tag{4.10}$$

for all $\delta < \delta_0$ and for all $\tilde{\lambda} \in \Lambda_h$.

Next, we use $\hat{y}_0$ to approximate the solution $\hat{y}_2(\xi, \tilde{\lambda}, \delta)$ of the nonautonomous equations (2.18). To this end, select $y_* = y_0(\xi_0) = y_2(\xi_0, \tilde{\lambda}, \delta)$ as above, and set $\Delta(\xi) = y_0 - y_2$. Then Δ satisfies the linear equation

$$\begin{aligned} \Delta' &= a_0(\tilde{\lambda}, \delta)\Delta + (a_0(\tilde{\lambda}, \delta) - a(\xi, \tilde{\lambda}, \delta))y_2(\xi, \tilde{\lambda}, \delta) \\ \Delta(\xi_0) &= 0. \end{aligned} \tag{4.11}$$

Recall that we have normalized $y_2(\xi_0, \tilde{\lambda}, \delta) = y_*$ to be a unit vector. By (4.7) the eigenvalues of a are exponentially close to those of a_0. Furthermore, we have that $\mathrm{Re}\, \mu_1(\tilde{\lambda})d(\tilde{\lambda}) < C$ for some $C > 0$, so that

$$|y_2(\xi, \tilde{\lambda}, \delta)| \leq e^{\mathrm{Re}\, |\mu_1| d(\tilde{\lambda})/\delta^\sigma} \leq Ce^{C/\delta^\sigma}, \tag{4.12}$$

uniformly on the interval $[\xi_0, \xi_+]$ and for $\tilde{\lambda} \in \Lambda$. Hence, if $F(\xi) = (a - a_0)y_2$ is the "inhomogeneous" term in (4.11), (4.12) together with (4.7) gives the uniform bound on the interval $\xi \in [\xi_0, \xi_+]$,

$$|F(\xi)| \leq Ce^{-\sqrt{b}/\delta^s} e^{C/\delta^\sigma} \leq Ce^{-(R+r)/\delta^s},$$

for some $r > 0$ depending only on b and $s - \sigma > 0$

Now express the solution of (4.11) as

$$\Delta(\xi) = \int_{\xi_0}^{\xi} e^{(\xi - t)a_0(\tilde{\lambda}, \delta)} F(t)\, dt.$$

Let $V = [e_1, e_2, e_3, e_4]$ be the matrix of eigenvectors of a_0, and let D be the diagonal matrix with the eigenvalues μ_i on the main diagonal. We then have that $e^{\xi a_0} = Ve^{\xi D}V^{-1}$. We have that $||V|| \leq C\delta^q$ where $q = \beta + 2(s - 1)$, so that $||V^{-1}|| \leq C^{-1}\delta^{-q}$, uniformly for $\tilde{\lambda} \in \Lambda_h$. The exponent q is of indeterminate sign; the point is that both V and V^{-1} are at worst algebraically large. It then follows for $\xi \in [\xi_0, \xi_+]$ that

$$\begin{aligned} |\Delta(\xi)| &\leq C|\xi - \xi_0|\delta^{-|q|} e^{\mathrm{Re}\, \mu_1(\tilde{\lambda})/\delta^\sigma} e^{-(R+r)/\delta^s} \\ &\leq [C\delta^{-(|q|+\sigma)} e^{C/\delta^\sigma} e^{-r/\delta^s}] e^{-R/\delta^s}. \end{aligned}$$

Since $s > \sigma$, the coefficient in brackets in the above estimate can be made arbitrarily small for all $\delta \in \Lambda_h$. It therefore follows that there exists $\delta(\Lambda_h) > 0$ such that

$$\rho[\hat{y}_2(\xi_+, \tilde{\lambda}, \delta), \hat{y}_0(\xi_+)] < e^{-R/\delta^s}, \tag{4.13}$$

for all $\delta < \delta_0(\Lambda_h)$ and all $\tilde{\lambda} \in \Lambda_h$.

A contradiction is now easily obtained by combining (4.10) and (4.13) to see that $\hat{y}_2(\xi_+, \tilde{\lambda}, \delta)$ is $\mathbf{O}(e^{-R/\delta^\sigma})$ close to $\hat{e}_1(\tilde{\lambda})$, uniformly for $\tilde{\lambda} \in \Lambda_h$ and for $\delta < \delta(\Lambda_h)$. If $|\lambda| \leq H$ with $\tilde{\lambda} = (\lambda, \pm\sqrt{\lambda - \lambda_*(\delta)})$, it follows that $(\hat{y}_2(\xi_+, \tilde{\lambda}, \delta), \xi_+) \in \Omega_1$, since $\xi_+ \leq -1/\delta^s < T_L$ so that $V_0(\tilde{\lambda}) = \Omega_1 \cap \{\xi = \xi_+\}$ is a small but $\mathbf{O}(1)$ neighborhood of $\hat{e}_1(\tilde{\lambda})$. Thus by Corollary 3.8, $\hat{y}_2 \to \hat{e}_1(\tilde{\lambda})$, which contradicts Lemma 4.2.

When $|\lambda| \geq H$ we similarly obtain a contradiction by applying the above estimates for $\hat{y}_2(\xi_+, \tilde{\lambda}, \delta)$ to verify that the representation (3.11) of (2.18) in the local coordinates (ζ_1, ρ, γ) lies in the neighborhood $V_o(\omega)$ at $\xi = \xi_+$. A contradiction is then obtained from (i) of Lemma 3.2. □

The estimate of Lemma 4.4 can be sharpened along the left slow field. In particular, we show below that $\hat{y}_2$ is exponentially close to $\hat{e}_2$ for $\xi < -2/\delta^s$. This is most easily seen when $\tilde{\lambda}$ is on the "standard" sheet S_1 of the Riemann surface $\mathcal{R}$ where Re $\mu_2 > 0 >$ Re μ_3. In that case, $\hat{e}_2(\lambda, \delta)$ is an attracting rest point for the restriction of the asymptotic projectivized flow $\hat{y}' = \hat{a}_0(\lambda, \delta, \hat{y})$ to the slow subspace $\hat{\Sigma}$. It then easily follows from the exponential estimate of Lemma 4.4 and the defining condition of $\hat{y}_2$ at $-\infty$ that the solution $\hat{y}_2(\xi, \lambda, \delta)$ of the nonautonomous equations (2.18) remains in a small neighborhood of $\hat{e}_2(\lambda, \delta)$ along the entire left slow field, $\xi \leq -2/\delta^s$. With a little care, this can be improved to yield an exponential estimate. This argument is *not* valid on the second sheet of the Riemann surface $\mathcal{R}$, since Re $\mu_3 > 0 >$ Re μ_2 here, and $\hat{e}_2$ is now a repeller for the slow subsystem. A different argument must therefore be provided in this case. It is here that we make crucial use of the fast Gap Lemma decay described in Chapter 2, section 2.

Here, the main issue is to obtain an estimate for $\tilde{\lambda}$ near the origin in the region with negative spectral gap. Uniform, large $|\lambda|$ behavior for Re $\lambda > 0$ can be derived using the scaling arguments of the previous lemma together with the attraction property of $\hat{e}_2(\lambda, \delta)$ for such λ. We therefore omit the details of this aspect of the estimate in the following lemma.

LEMMA 4.5. *Suppose that $0 < \tilde{R} < R$, where R is as in Lemma 4.4. The estimate*

$$\rho[\hat{y}_2(\xi, \tilde{\lambda}, \delta), \hat{e}_2(\tilde{\lambda}, \delta)] < e^{-\tilde{R}/\delta^\sigma}$$

holds for all sufficiently small δ, for all $\xi \leq -2/\delta^s$, and for all $\tilde{\lambda} \in \Lambda_h$.

PROOF. Fix $\xi_o < -2/\delta^s$ and suppose that the lemma is false at ξ_o. Consider the subspace of solutions $\Phi(\xi, \tilde{\lambda}, \delta)$ of Lemma 2.5 which tends as $\xi \to -\infty$ to the subspace $\Phi(\tilde{\lambda}, \delta)$ defined in (2.8). We then have that

$$\Phi(\xi, \tilde{\lambda}, \delta) = \text{span}\{y_1(\xi, \tilde{\lambda}, \delta), y_2(\xi, \tilde{\lambda}, \delta)\},$$

where $y_1 \in \hat{y}_1$ is uniquely determined up to scalar multiples by Lemma 2.2, and $y_2 \in \hat{y}_2$ is uniquely determined in Φ up to scalar multiples for $\tilde{\lambda} \in \Lambda_h$ by Lemma 4.2. Since Re $\mu_1(\tilde{\lambda})$ is strongly separated from the rest of the spectrum of a_o, it follows from Lemma 2.5 in **[14]** that there is a unique choice of $y_1 \in \hat{y}_1$ satisfying the asymptotic decay estimate

$$y_1(\xi, \tilde{\lambda}, \delta) = e^{\mu_1(\tilde{\lambda})(\xi - \xi_o)}[e_1(\tilde{\lambda}) + \mathbf{O}(e^{\tilde{\kappa}\xi})]$$

as $\xi \to -\infty$. Here $\tilde{\kappa} < \kappa$, where κ satisfies (2.9). Now select $y_2 \in \hat{y}_2$ so that $||y_2(\xi_o, \tilde{\lambda}, \delta)|| = 1$; there exist L_i such that $y_2(\xi_o, \tilde{\lambda}, \delta) = \Sigma_{i=1}^4 L_i e_i(\tilde{\lambda}, \delta)$. It follows from Lemma 4.4 that $|L_1|, |L_4| \leq Ce^{-R/\delta^\sigma}$, so that

$$y_2(\xi_o, \tilde{\lambda}, \delta) = L_2 e_2(\tilde{\lambda}, \delta) + L_3 e_3(\tilde{\lambda}, \delta) + \mathbf{O}(e^{-R/\delta^\sigma}),$$

where $C > 0$ is some constant depending only on the metric ρ on $\mathbb{CP}^3$. At least one of the coefficients L_2, L_3 must be at least $\mathbf{O}(1)$ since $y_2(\xi_o, \tilde{\lambda}, \delta)$ is a unit vector. The strategy is to show that the exponential decay rate of the 2-form $\eta = y_1 \wedge y_2$ as

$\xi \to -\infty$ then must occur at a slower rate than the Gap Lemma rate, as specified in (2.12).

Since we are assuming that $\rho[\hat{y}_2, \hat{e}_2] \geq e^{-\tilde{R}/\delta^\sigma}$ at $\xi = \xi_0$ and $\tilde{R} < R$ it follows that there exists $C > 0$ depending only on $R - \tilde{R} > 0$ and the metric ρ, but not on δ, such that

$$|L_3| \geq Ce^{-\tilde{R}/\delta^\sigma}, \tag{4.14}$$

The equations for y_2 can be expressed as

$$y_2' = a_o(\tilde{\lambda}, \delta)y_2 + f(\xi), \tag{4.15}$$

where $f(\xi) = (a(\xi, \tilde{\lambda}, \delta) - a_o(\tilde{\lambda}, \delta))y_2(\xi, \tilde{\lambda}, \delta)$. Since $\hat{y}_2$ remains exponentially close to $\hat{\Sigma}$ for $\xi \leq \xi_o$ and $y_2(\xi_o, \tilde{\lambda}, \delta)$ is a unit vector, it follows that there is constant $C > 0$ independent of δ such that $||y_2(\xi, \tilde{\lambda}, \delta)|| < Ce^{C\delta|\xi - \xi_o|}$ for $\xi \leq \xi_o$, since $|\mu_{2,3}| = \mathbf{O}(\delta)$. On the second sheet of $\mathcal{R}$, y_2 has slow exponential growth at $-\infty$, but by Lemma 4.4, it cannot have fast exponential growth in this direction. Combining (4.7) with the growth bound for y_2, we finally obtain

$$||f(\xi)|| < Ce^{\tilde{R}\xi}$$

for $\xi \leq \xi_0$, where $0 < \tilde{R} < R$ is independent of δ. Thus the inhomogeneous term in (4.15) decays at a fast exponential rate.

Now let $w_i(\xi, \tilde{\lambda}, \delta) = e^{\mu_i(\tilde{\lambda}, \delta)(\xi - \xi_0)}e_i(\tilde{\lambda}, \delta)$ be four independent solutions of the linear equation $w' = a_o(\tilde{\lambda}, \delta)w$, and express the solution y_2 of (4.15) as $y_2 = \sum_{i=1}^4 (L_i + a_i(\xi))w_i(\xi, \tilde{\lambda}, \delta)$. By Lemma 4.4, $\hat{y}_2$ is exponentially close to the slow subspace over the left slow field. Thus the crucial term is $a_3(\xi)$. The variation of parameters expression for a_3 is

$$a_3(\xi) = \det[e_1, e_2, e_3, e_4]^{-1} \int_{\xi_0}^{\xi} e^{(\mu_1 + \mu_2 + \mu_4)(\xi - \xi_o)} \det[e_1, e_2, f, e_4]\, d\xi,$$

since $a_3(\xi_0) = 0$ and a_o has trace zero. Now $\mu_1 + \mu_2 + \mu_4 = -\mu_3 = \mathbf{O}(\delta)$, and it follows from the above expression for $a_3(\xi)$ and the fast exponential decay estimate for $f(\xi)$ that

$$\begin{aligned} |a_3(\xi)| &\leq \int_{\xi_o}^{\xi} \mathbf{O}(e^{\tilde{R}\xi})\, d\xi \\ &\leq Ce^{-\tilde{R}/\delta^s} \end{aligned}$$

for $\xi \leq \xi_0 \leq -2/\delta^s$ and all sufficiently small δ. Here $\tilde{R}$ has been modified by an $\mathbf{O}(\delta)$ amount to absorb the slow growth of the exponential factor in the integral. It now follows from (4.14) the above, and $\sigma < s$ that $|L_3 + a_3(\xi)| \geq Ce^{-\tilde{R}/\delta^\sigma}/2$ for all $\xi < \xi_o$, so that this coefficient is bounded away from zero for $\xi < \xi_o$. We then have that

$$\begin{aligned} \eta(\xi, \tilde{\lambda}, \delta) &= y_1(\xi, \tilde{\lambda}, \delta) \wedge y_2(\xi, \tilde{\lambda}, \delta) \\ &= e^{(\mu_1 + \mu_2)(\xi - \xi_o)} \cdot \\ &\quad [\Sigma_{i=2}^4 (L_i + a_i(\xi))e_1(\tilde{\lambda}) \wedge e_i(\tilde{\lambda}, \delta)e^{(\mu_i - \mu_2)(\xi - \xi_o)} + \mathbf{O}(e^{\tilde{\kappa}\xi})]. \end{aligned}$$

Since the coefficient $(L_3 + a_3)$ is bounded away from zero for all $\xi < \xi_o$, the exponential decay rate inside the brackets can be no faster than $\mu_3 - \mu_2 = \mathbf{O}(\delta)$. This contradicts the decay rate (2.12) for η. □

4.3. Tracking the Slow Solution over the Fast Field

Next, we approximate y_2 over the fast field, which we define to be the interval $|\xi| < 2/\delta^s$, where $0 < s < 1$; s will need to satisfy some additional conditions specified below. Lemmas 4.4 and 4.5 of the previous section provide estimates for $\hat{y}_2$ at $\pm 2/\delta^s$, however, in order to apply these estimates to a solution y_2 of the original linear equations (2.5), we need to pick a normalization of the solution. This can only be done for one value of ξ, whereas two normalizations are required in order to use the lemmas to approximate y_2 at $\pm 2/\delta^s$. The following lemma is used to address this difficulty.

LEMMA 4.6. *Let $y_*(\xi, \tilde{\lambda}, \delta) = (u_*, p_*, v_*, q_*)(\xi, \tilde{\lambda}, \delta) \in \hat{y}_2(\xi, \tilde{\lambda}, \delta)$ be normalized so that*

$$\max_{|\xi| \leq 2/\delta^s} \{|u_*(\xi, \tilde{\lambda}, \delta)| + |p_*(\xi, \tilde{\lambda}, \delta)|\} = 1. \tag{4.16}$$

Let

$$M(\delta, \tilde{\lambda}) = \max_{|\xi| \leq 2/\delta^s} \{|v_*(\xi, \tilde{\lambda}, \delta)| + |q_*(\xi, \tilde{\lambda}, \delta)|\};$$

then $M(\delta, \tilde{\lambda}) \leq M$ for some $M > 0$ for all $\tilde{\lambda} \in \Lambda_h$ and $\delta < \delta(\Lambda_h)$.

PROOF. The proof of this lemma is very similar to that of Lemma 5.7 of [**13**]; we will therefore only sketch the proof. If the theorem were false, there would exist sequences $\delta_n \to 0$, $\tilde{\lambda}_n$, and $|\xi_n| \leq 2/\delta^s$ such that $M(\delta_n, \tilde{\lambda}_n) \to \infty$. First assume that $|\lambda_n| \leq H$ for all n, where λ_n is the projection of $\tilde{\lambda}_n$ on S_1 Without loss of generality, $\tilde{\lambda}_n$ converges to a finite limit $\tilde{\lambda}$ with $|\lambda| \leq H$, We may assume then that $\tilde{\lambda}_n = \tilde{\lambda}_0$ for all n with $|\lambda| \leq H$. There are two distinct cases: (i) the sequence ξ_n has no finite limit point, and (ii) ξ_n has a finite limit point ξ_o. In case (i), we also distinguish between $|\lambda| \leq H$ and $|\lambda| \geq H$. In the first case, by passing to a suitable subsequence it can be assumed that either $\xi_n >> T_R$ or $\xi_n << T_L$ for all n, where T_L, T_R as in the construction of the tubes $\Omega_{1,4}$ of Chapter 3. In either case, it follows from the fact that $M(\delta_n, \tilde{\lambda}) \to \infty$ while the slow components of y_* remain uniformly bounded that $\hat{y}_*(\xi_n, \tilde{\lambda}_0, \delta_n)$ approaches the (projectivization) of the span of $e_{1,4}$ as $n \to \infty$. It would then follow that $\hat{y}_2(\xi_n \pm T_*, \tilde{\lambda}, \delta_n)$ enters at least one of the tubes $\Omega_{1,4}$ in at least one time direction, after some large but $\mathbf{O}(1)$ time interval from the time $\pm T_*$, for all sufficiently large n. The details are similar to and simpler than those of the proof of Lemma 4.4, and we omit the details; see also [**13**].

In case (ii), the estimate is a bit more delicate, since ξ_n remains in some $\mathbf{O}(1)$ interval about $\xi = 0$. Without loss of generality, it can be assumed that ξ_n converges to a finite limit ξ_0. Since $M_n \to \infty$, it follows from the normalization chosen in the statement of the lemma that y_*/M_n has (slow) components u_*, p_* tending to zero uniformly on compact intervals, and fast components v_*, q_* normalized so that $\max\{|v|, |q|\} = 1$ at $\xi = \xi_n \to \xi_o$. It again follows that there is some uniform $\mathbf{O}(1)$ time T_* such that $\xi_n \pm T_* \notin [T_L, T_R]$ and $\hat{y}_2(\xi_n \pm T_*)$ is an $\mathbf{O}(1)$ distance from the projectivized slow subspace $\hat{\Sigma}$. Once again the projectivized solution $\hat{y}_2$ must enter at least one of the tubes $\Omega_{1,4}$ at time $\xi_n \pm T_{**}$ in $\mathbf{O}(1)$ time $\pm T_{**}$, where $T_{**} > T_*$.

If $\tilde{\lambda}_o = \lambda_o$ with $|\lambda_o| \geq H$ we obtain a contradiction from Lemma 3.2 and Corollary 3.11. In this case, we use the equations (3.11) for the local coordinate representation (ζ_1, ρ, γ) of the solution and an argument similar to the above to see that there is a uniform $\mathbf{O}(1)$ time $T_* > 0$ such that the solution $\hat{y}_2(\xi_n \pm T_*, \lambda_o, \delta_n)$ enters at least one of the neighborhoods $V_0(\omega)$ or $W_0(\omega)$ of Lemma 3.2 and Corollary 3.11. A contradiction is then obtained as before. □

Next, we show that the slow solution y_2 is determined to leading order by the solution of the NLEP described in Chapter 1. We first consider the case of bounded $\tilde{\lambda} \in \Lambda_h$, in which $\tilde{\lambda} \in \Lambda_h$ is assumed to satisfy $|\lambda| \leq H$. Here $H > 0$ is a large, $\mathbf{O}(1)$ constant which satisfies the requirements of Lemma 3.2 together with an additional condition specified below stemming from the slow analysis. The main estimate in this regard is given in Theorem 4.11. We then consider the "large λ" case, wherein $\lambda \in \Lambda_h$ satisfies $|\lambda| \geq H$. This is also a somewhat delicate situation to analyze, since there is now a small parameter δ and an independent, large parameter λ present in the same equation (recall Lemma 3.2). It is most convenient to present this result after the analysis of the case where $|\lambda| \leq H$.

LEMMA 4.7. *Let $r = (1-\beta) - s$ and suppose that $r > 0$. Also, suppose that $\tilde{\lambda} \in \Lambda \cap \{\lambda \in S_1 : |\lambda| \leq H\}$ and that $y_2(\xi, \tilde{\lambda}, \delta)$ is chosen so that*

$$y_2(-2/\delta^s, \tilde{\lambda}, \delta) = e_2(\tilde{\lambda}, \delta) + \mathbf{O}(e^{-\tilde{R}/\delta^\sigma}). \tag{4.17}$$

Then $|y_2(\xi, \tilde{\lambda}, \delta)|$ is uniformly bounded for $|\xi| \leq 2/\delta^s$ and for all sufficiently small $\delta > 0$, and there exist $R_2(\tilde{\lambda}, \delta)$ and $R_3(\tilde{\lambda}, \delta)$ such that

$$\begin{aligned} y_2(+2/\delta^s, \tilde{\lambda}, \delta) &= R_2(\tilde{\lambda}, \delta) e_2(\tilde{\lambda}, \delta) + R_3(\tilde{\lambda}, \delta) e_3(\tilde{\lambda}, \delta) + \mathbf{O}(e^{-\tilde{R}/\delta^\sigma}), \\ R_2(\tilde{\lambda}, \delta) + R_3(\tilde{\lambda}, \delta) &= 1 + \mathbf{O}(\delta^r) \\ R_2(\tilde{\lambda}, \delta) - R_3(\tilde{\lambda}, \delta) &= 1 + \mathbf{O}(\delta^{r-\beta}). \end{aligned} \tag{4.18}$$

PROOF. Let $y_*(\xi, \tilde{\lambda}, \delta) \in \hat{y}_2(\xi, \tilde{\lambda}, \delta)$ be normalized as in (4.16). By Lemma 4.6, all components of y_* remain uniformly bounded over the fast field $|\xi| \leq 2/\delta^s$. It also follows from Lemma 3.1 that the various coefficients in of the equations in (2.2) also remain uniformly bounded over the fast field for all small $\delta > 0$. Thus, if $\xi_1, \xi_2 \in [-2/\delta^s, +2/\delta^s]$, it follows upon integration of the first two equations in (2.2) that

$$\begin{aligned} |u_*(\xi_1, \tilde{\lambda}, \delta) - u_*(\xi_2, \tilde{\lambda}, \delta)| &< \mathbf{O}(\delta^r) \\ |p_*(\xi_1, \tilde{\lambda}, \delta) - p_*(\xi_2, \tilde{\lambda}, \delta)| &< \mathbf{O}(\delta^r) \end{aligned} \tag{4.19}$$

for all $|\xi_1|, |\xi_2| \leq 2/\delta^s$. Suppose that the maximum in (4.16) occurs at ξ_1. Setting $\xi_2 = -2/\delta^s$ in (4.19) then yields

$$|u_*(-2/\delta^s, \tilde{\lambda}, \delta)| + |p_*(-2/\delta^s, \tilde{\lambda}, \delta)| = 1 + \mathbf{O}(\delta^r). \tag{4.20}$$

It then follows from Lemma 4.4 and the above that there is an $\mathbf{O}(1)$ constant $\gamma \in \mathbb{C}$ such that

$$y_*(-2/\delta^s, \tilde{\lambda}, \delta) = \gamma e_2(\tilde{\lambda}, \delta) + \mathbf{O}(e^{-\tilde{R}/\delta^\sigma}). \tag{4.21}$$

More explicitly, we see from Lemma 2.1, (4.20), and the above that

$$|\gamma| = \frac{1 + \mathbf{O}(\delta^r)}{1 + \delta^\beta |\sqrt{b\lambda + \delta^{2-\beta} a}|}. \tag{4.22}$$

When $\beta > 0$, $\gamma = 1 + \mathbf{O}(\delta^\beta + \delta^r)$, and when $\beta = 0$ it is still an $\mathbf{O}(1)$ constant. Hence, we find that $y_2 = \gamma^{-1} y_*$, when y_2 is normalized as in (4.17).

Since $y_2 = \gamma^{-1} y_*$ for all ξ, we obtain from Lemma 4.6 that all components of y_2 are uniformly bounded for all $|\xi| \le 2/\delta^s$ and for all sufficiently small δ. It therefore follows that an estimate of the form (4.19) also holds for the u and p components of y_2 over the fast field. The relations between R_2 and R_3 in (4.18) then follow from using the explicit expressions for $e_{2,3}$ in Lemma 2.1, the normalization (4.17) of y_2 at $-2/\delta^s$, and (4.19) with $\xi_1 = -2/\delta^s$ and $\xi_2 = +2/\delta^s$. □

REMARK 4.8. The coefficients $R_{2,3}$ can be algebraically large in δ even though y_2 is bounded. This is because $e_2 \wedge e_3 = \mathbf{O}(\delta^\beta)$.

LEMMA 4.9. *Suppose that $\tilde{\lambda} \in \Lambda_h \cap \{\lambda \in S_1 : |\lambda| \le H\}$ and that $v(\xi, \tilde{\lambda}, \delta)$ is the v-component of $y_2(\xi, \tilde{\lambda}, \delta)$, where $y_2 = (u, p, v, q)(\xi, \tilde{\lambda}, \delta)$ is chosen as in (4.17). Let I_δ be the integral operator*

$$I_\delta(v, \tilde{\lambda}, \delta) = \frac{\delta^{1-2\beta}}{\pm 2[b(\lambda - \lambda_*(\delta))]^{1/2}} \int_{-2/\delta^s}^{2/\delta^s} [v_o(\xi, \delta)^2 + 2u_o(\xi, \delta) v_o(\xi, \delta) v(\xi, \lambda, \delta)]\, d\xi,$$

where the $\sqrt{\ }$ is the standard branch of the square root with branch cut along the negative real axis. The sign in front of the square root in I_δ is determined by the $\pm$ sign in the local coordinate representation $\tilde{\lambda} = (\lambda, \pm\sqrt{\lambda - \lambda_*(\delta)})$ of the point $\tilde{\lambda} \in \Lambda$, where λ is on the first sheet of $\mathcal{R}$.

Let $s = (1-\beta)/3$. There is a continuous function $m(\beta)$ (see (4.28) below) such that $m(\beta) > 1 - 2\beta$ for $\beta \in [1/2, 1)$ and $m(\beta) > 0$ for $\beta \in [0, 1/2]$, and such that

$$\begin{aligned} R_2(\tilde{\lambda}, \delta) &= 1 + I_\delta(v, \tilde{\lambda}, \delta) + \mathbf{O}(\delta^{m(\beta)}) \\ R_3(\tilde{\lambda}, \delta) &= -I_\delta(v, \tilde{\lambda}, \delta) + \mathbf{O}(\delta^{m(\beta)}), \end{aligned} \tag{4.23}$$

for all sufficiently small $\delta > 0$.

PROOF. From Lemma 2.1 the p component of $e_2(\tilde{\lambda}, \delta)$ is $\pm\delta^\beta\sqrt{b(\lambda - \lambda_*(\delta))}$, which is of order δ^β. It follows from Lemmas 3.1, 4.5, 4.6 and the normalization (4.17) that all terms in the p equation in(2.2) remain uniformly bounded over the fast field, so that and an integration of the second equation in (2.2) yields the sharper estimate that $|p| = \mathbf{O}(\delta^\beta) + \mathbf{O}(\delta^r)$ for $|\xi| \le 2/\delta^s$. Here we have used the relation $r = 1 - \beta - s$. Since the u component of e_2 is 1, it then follows from (4.17) and a similar argument applied to an integration of the first equation in (2.2) over the fast field that

$$u(\xi, \tilde{\lambda}, \delta) = 1 + \mathbf{O}(\delta^{r+\beta} + \delta^{2r}), \tag{4.24}$$

for $|\xi| \le 2/\delta^s$. However, we also have the expression (4.18) for y_2 at $2/\delta^s$. Since the u-component of $e_{2,3}(\tilde{\lambda}, \delta)$ is 1, it then follows from (4.24) and the first equation in (4.18) that

$$R_2(\tilde{\lambda}, \delta) + R_3(\tilde{\lambda}, \delta) = 1 + \mathbf{O}(\delta^{r+\beta} + \delta^{2r}). \tag{4.25}$$

Here, we have absorbed the exponential errors into the algebraic terms. This is a sharper estimate than (4.18).

Next, we use (4.24) in an integration of the p-equation in (2.2) over the fast field. Since the coefficient $v_o(\xi, \delta)$ decays at a fast rate $\mathbf{O}(e^{-\sqrt{b}|\xi|})$, it follows that the p equation in (2.2) can be integrated to yield an estimate for the change in p over the fast field,

$$\Delta p \;=\; \delta^{1-\beta} \int_{-2/\delta^s}^{2/\delta^s} [v_o(\xi, \delta)^2 + 2u_0(\xi, \delta) v_o(\xi, \delta) v(\xi, \tilde{\lambda}, \delta)]\, d\xi$$

$$+ \quad \mathbf{O}(\delta^{r+1-s} + \delta^{3r} + \delta^{2\beta+r}) \tag{4.26}$$

where $\Delta p = p(2/\delta^s, \tilde{\lambda}, \delta) - p(-2/\delta^s, \tilde{\lambda}, \delta)$, and where the $\mathbf{O}(\delta^{2\beta+r})$ term arises from the two higher order constant terms in the coefficient γ_u in the p' equation in (2.2).

The change in p can also be calculated from the difference in the p component of y_2 in the expressions (4.17) and (4.18). Using the explicit expressions for $e_{2,3}$ in Lemma 2.1, we then find that

$$\Delta p = \pm\delta^\beta\sqrt{b(\lambda - \lambda_*(\delta))}(R_2(\tilde{\lambda}, \delta) - R_3(\tilde{\lambda}, \delta) - 1) + \mathbf{O}(e^{-\tilde{R}/\delta^\sigma}),$$

where the sign of the square root is determined as in the statement of the lemma. Equating this expression with (4.26) and dividing by δ^β yields the equation

$$(R_2(\tilde{\lambda}, \delta) - R_3(\tilde{\lambda}, \delta) - 1) = I_\delta(v, \tilde{\lambda}, \delta) + \mathbf{O}(\delta^{2r} + \delta^{3r-\beta} + \delta^{\beta+r}), \tag{4.27}$$

where we have used the relation $r = 1 - \beta - s$, and all exponentially small errors have been absorbed into the algebraic terms. For $\beta < 1/2$, the integral term itself has a coefficient of higher order, so that the relative size of $1 - 2\beta$ and the error term is irrelevant, provided that all the powers of δ in the latter, $2r, 3r - \beta$, and $\beta + r$ are positive. For $1/2 \leq \beta \leq 1$, the integral term becomes the leading order term, and we require that these exponents are all larger that $1 - 2\beta$.

Finally, set $s = (1 - \beta)/3$, so that $r = 2(1 - \beta)/3$. The exponents in the error term in (4.27) are then all larger than $m(\beta)$, where

$$m(\beta) = \min\{4(1 - \beta)/3, (2 - 3\beta), (2 + \beta)/3\}. \tag{4.28}$$

This function is clearly positive for $0 \leq \beta \leq 1/2$. It is also easily verified that $1 - 2\beta < m(\beta)$ for $1/2 < \beta < 1$, as required.

The statement of the lemma is finally obtained by combining (4.27) and (4.25). □

REMARK 4.10. We have chosen $s = (1 - \beta)/3$ for convenience, and there is no particular significance to our choice other than it gives some error bound. It might be interesting to try to optimize the choice of s as a function of β to obtain the sharpest possible error estimates. This information might be of some use in numerical calculations.

We can now state the main result of this chapter. To this end, let $\tilde{\lambda} \in \Lambda_h$ be expressed in local coordinates as $\tilde{\lambda} = (\lambda, \pm\sqrt{\lambda - \lambda_*(\delta)})$, with the sign of the square root determined as in Lemma 4.9. Since $\tilde{\lambda} \in \Lambda$, λ is *not* an eigenvalue of the homogeneous equation associated to the inhomogeneous differential equation in (1.13). By standard Fredholm theory, there exists a unique, bounded solution $v_*(\xi, \lambda)$ of the inhomogeneous differential equation in (1.13). In fact, it is easily seen that v_* has exponential decay at infinity (e.g., see below). Finally, define the integral operator,

$$I_*(v_*, \tilde{\lambda}, \delta) = \frac{\delta^{1-2\beta}}{\pm 2\sqrt{b(\lambda - \lambda_*(\delta))}} \int_{-\infty}^{\infty} [S(\xi)^2 + 2\bar{c}S(\xi)v_*(\xi, \tilde{\lambda})] \, d\xi. \tag{4.29}$$

THEOREM 4.11. . *Let $\tilde{\lambda} \in \Lambda_h \cap \{\lambda \in S_1 : |\lambda| \leq H\}$, and let v_*, and I_* be as in (4.29). Then the coefficients $R_2(\tilde{\lambda}, \delta), R_3(\tilde{\lambda}, \delta)$ in (4.18) for the solution $y_2(2/\delta^s, \tilde{\lambda}, \delta)$ of the perturbed equations (2.4) satisfy*

$$\begin{aligned} R_2(\tilde{\lambda}, \delta) &= 1 + I_*(v_*, \tilde{\lambda}, \delta) + \mathbf{O}(\delta^{\tilde{m}(\beta)}) \\ R_3(\tilde{\lambda}, \delta) &= -I_*(v_*, \tilde{\lambda}, \delta) + \mathbf{O}(\delta^{\tilde{m}(\beta)}), \end{aligned} \tag{4.30}$$

for all sufficiently small $\delta > 0$, *where* $\tilde{m}(\beta) = \min\{m(\beta), (5-8\beta)/3\}$, *and* $\tilde{m}(\beta) > 1 - 2\beta$ *for all* $\beta \in [0,1)$.

PROOF. Let $I_* = I_{*f} + I_{*s}$ where

$$\begin{aligned} I_{*f} &= \frac{\delta^{1-2\beta}}{\pm 2\sqrt{b(\lambda - \lambda_*(\delta))}} \int_{-2/\delta^s}^{2/\delta^s} [S(\xi)^2 + 2\bar{c}S(\xi)v_*(\xi, \tilde{\lambda})]\, d\xi \\ I_{*s} &= \frac{\delta^{1-2\beta}}{\pm 2\sqrt{b(\lambda - \lambda_*(\delta))}} \int_{|\xi| \geq 2/\delta^s} [S(\xi)^2 + 2\bar{c}S(\xi)v_*(\xi, \tilde{\lambda})]\, d\xi. \end{aligned}$$

It is clear (see below) that the integral I_{*s} over the slow fields is exponentially small in δ. Hence the main task is to estimate $|I_{*f} - I_\delta|$.

To this end, we first replace the perturbed wave $(u_o(\xi,\delta), v_o(\xi,\delta))$ by the fast limit $(\bar{c}, S(\xi))$ in the integrand of I_δ. By the (3.1) of Lemma 3.1, this incurs an error of order $\mathbf{O}(\delta^{1-\beta})$ in the integrand of I_δ, since by Lemma 4.6, all components of y_2, and in particular $v(\xi, \tilde{\lambda}, \delta)$, are uniformly bounded over $|\xi| \leq 2/\delta^s$. It therefore follows that the error in I_δ is of order $\delta^{1-2\beta}\delta^{1-\beta-s} = \delta^{2-3\beta-s}$ Setting $s = (1-\beta)/3$ then yields the estimate

$$|I_{*f}(v, \tilde{\lambda}, \delta) - I_\delta(v, \tilde{\lambda}, \delta)| = \mathbf{O}(\delta^{(5-8/\beta)/3}).$$

Next, we need to approximate $|v - v_*|$. The equations satisfied by (u, v) are (1.9), while v_* satisfies the differential equation in (1.13). By Lemma 3.1 the coefficients of (1.9) are uniformly approximated by those of (1.13) to within an error of $\mathbf{O}(\delta^{1-\beta})$ on the interval $|\xi| \leq 2/\delta^s$. Using this together with the approximation (4.24) for u on $|\xi| \leq 2/\delta^s$, the equation (1.9) for v can then be expressed as

$$v'' + [2\bar{c}S(\xi) + b(1+\lambda)]v = -S(\xi)^2 + \mathbf{O}(\delta^{1-\beta} + \delta^{2r} + \delta^{r+\beta}) \tag{4.31}$$

over $|\xi| \leq 2/\delta^s$.

This approximation would appear to be valid only over the fast field, $|\xi| \leq 2/\delta^s$. However, y_2 has at worst slow exponential growth at ∞ of order $e^{C\delta|\xi|}$ for some constant $C > 0$, while the coefficients $v_o(\xi, \delta)$ and $S(\xi)$ have rapid exponential decay at $\pm\infty$ of order $\mathbf{O}(e^{-\sqrt{b}|\xi|})$. Hence, every term in the equation in (1.9) for v with a factor of $v_o(\xi, \delta)$ is exponentially small in δ for $|\xi| \geq 2/\delta^s$, and these terms decay at a rapid exponential rate as $|\xi| \to \infty$. Hence, the equation (4.31) is actually uniformly valid on $\mathbb{R}$, since the exponential error terms for $|\xi| \geq 2/\delta^s$ are dominated by the algebraic error terms over the slow field.

Next, we construct the Green's function $G(\xi, y, \lambda)$ for the operator

$$L(\lambda)v = v'' + [2\bar{c}S(\xi) - b(1+\lambda)]v.$$

For $\lambda \neq 0, \lambda_1$ and $\operatorname{Re}\lambda > -1/2$, there exist solutions $w_\pm(\xi, \lambda)$ of $Lw = 0$ satisfying the exponential estimates

$$\begin{aligned} w_-(\xi, \lambda) &= e^{\sqrt{b(1+\lambda)}\xi}(1 + \mathbf{O}(e^{-\sqrt{b}|\xi|})) \quad \xi \to -\infty \\ w_+(\xi, \lambda) &= e^{-\sqrt{b(1+\lambda)}\xi}(1 + \mathbf{O}(e^{-\sqrt{b}|\xi|})) \quad \xi \to +\infty \end{aligned} \tag{4.32}$$

(see e.g. Lemma 2.6 in [**14**]). Note that the Wronskian $W = W(\lambda)$ of w_-, w_+ is the Evans function for the Fisher wave. Thus $W(\lambda)$ is constant in ξ and $W(\lambda)$ is bounded away from zero for $\tilde{\lambda} \in \Lambda$ since Λ is disjoint from the eigenvalues of the

Fisher wave. The Green's function for $L(\lambda)$ is then

$$G(\xi, y, \lambda) = \begin{cases} w_-(\xi, \lambda)w_+(y, \lambda)/W(\lambda) & (\xi < y) \\ w_+(\xi, \lambda)w_-(y, \lambda)/W(\lambda) & (y < \xi) \end{cases} . \tag{4.33}$$

Since $S(\xi)$ vanishes at a fast exponential rate $\mathbf{O}(e^{-\sqrt{b}|\xi|})$ and v_* is the *bounded* solution of $L(\lambda)v_* = -S(\xi)^2$, it follows that

$$v_*(\xi, \lambda) = -\int_{-\infty}^{\infty} G(\xi, y, \lambda)S(y)^2\, dy. \tag{4.34}$$

It then follows from this representation formula together with the asymptotic estimates (4.32) and L'Hospital's rule that the solution v_* and its derivatives satisfy a rapid exponential decay estimate over the slow fields, $|v_*(\xi, \tilde{\lambda})| \leq Ke^{-\sqrt{b(1+\lambda)}|\xi|}$. Since v satisfies (4.31) over the whole line, and, as noted above, the entire right hand side decays at a fast exponential rate for large $|\xi|$, it similarly follows that $|v(\xi, \tilde{\lambda}, \delta)|$ also decays at a rapid exponential rate at $\pm\infty$.

Let $\Delta = v - v_*$. It follows from the above estimates for v and v_* over the slow fields $|\xi| \geq 2/\delta^s$, together with the equation (4.31) for v over the fast field, that Δ satisfies the equation $L(\lambda)\Delta = \mathbf{O}(\delta^{1-\beta} + \delta^{2r} + \delta^{1-s})$, uniformly for all ξ. Here, we have absorbed the exponentially small contributions to the error over the slow fields into the algebraic terms. It again follows from the Green's function representation that $|\Delta(\xi, \lambda, \delta)| \leq \mathbf{O}(\delta^{1-\beta} + \delta^{2r} + \delta^{1-s})$, uniformly for all ξ. Replacing v by v_* in the integral in (4.23) induces an error of at most $\mathbf{O}(\delta^{1-\beta-s} + \delta^{2r-s} + \delta^{1-2s})$. Using the values $s = (1-\beta)/3$ and $r = 2(1-\beta)/3$, we find the each of the errors these terms induce in the integral are at most $\mathbf{O}(\delta^{2(1-\beta)/3})$. The net error induced by this is therefore $\delta^{1-2\beta}\delta^{2(1-\beta)/3} = \delta^{(5-8\beta)/3}$.

The fast exponential decay of $v_*(\xi, \lambda, \delta)$ and of $v_0(\xi, \delta)$ permit the integral in (4.23) over the slow field to be extended to the entire real line while inducing an error which is at most exponentially small in δ. This completes the proof. $\square$

In order to rule out the possibility of eigenvalues $\lambda \in \Lambda_h$ which become unbounded as $\delta \to 0$, we need to supplement Lemma 3.2 with an analysis of the behavior of $y_2(\xi, \tilde{\lambda}, \delta)$ for $\lambda \in \Lambda_h \cap \{|\lambda| \geq H\}$. The analysis in the Appendix of[**6**] provides analogous estimates for the NLEP for all fixed scalings of λ in δ as $\delta \to 0$; however, as in the case of Lemma 3.2, we require an analysis which applies to the subset $|\lambda| \geq H$ of Λ_h, uniformly for small $\delta > 0$. To this end, we first prove a preliminary lemma.

LEMMA 4.12. *Suppose that $y_*(\xi, \lambda, \delta) \in \hat{y}_2(\xi, \lambda, \delta)$ is a solution of (2.4) satisfying $|y_*(\xi, \lambda, \delta)| \leq M$ for all $|\xi| \leq 2/\delta^s + 1$. There exist $\delta_* > 0$ and an $\mathbf{O}(1)$ constant $K > 0$ such that $|v_*(\xi, \lambda, \delta)| \leq KM|\lambda|^{-1}$ for all $|\xi| \leq 2/\delta^s$, $\delta < \delta_0$, and $\lambda \in \Lambda_h \cap S_1$.*

PROOF. In the following, $K > 0$ will denote a generic $\mathbf{O}(1)$ constant that is uniformly bounded both from above and from below for all λ and δ as in the statement of the Lemma. It follows from Lemma 4.4 and the bound assumed for y_* over the fast field that there is exists $K > 0$ such that for $2/\delta^s \leq |\xi| \leq 1 + 2/\delta^s$, we have that

$$|v_*|, |q_*| \leq Ke^{-R/\delta^\sigma} \max\{|u_*|, |p_*|\} \leq KMe^{-R/\delta^\sigma},$$

for all $\lambda \in \Lambda_h$. This follows from the fact that $\rho[\hat{y}_2, \Sigma] \leq e^{-R/\delta^\sigma}$. Thus K in the above depends only on the metric ρ.

Let $\varphi(\xi)$ be a test function which satisfies $\varphi(\xi) = 1$ for $|\xi| \le 2/\delta^s$ and $\varphi(\xi) = 0$ for $|\xi| \ge 2/\delta^s + 1$, and let $w = \varphi v_*$. Thus w satisfies the equation

$$w'' + [2u_o v_o - b(1+\lambda)]w = -\varphi v_0^2 u + 2\varphi' q_* + \varphi'' v_* \equiv F(\xi). \tag{4.35}$$

Since the support of φ', φ'' is on $2/\delta^s \le |\xi| \le 1 + 2/\delta^s$, it follows that $|F(\xi)| \le KM$ for all ξ. (We need a cutoff function here to control the exponential growth of u over the slow field. For very large $|\lambda|$, this growth rate will exceed the decay rate of v_o^2.)

Let $w_o(\xi) = 2u_o(\xi,\delta)v_0(\xi,\delta)$ and $\omega = [b(1+\lambda)]^{1/2}$; then the associated homogeneous equation for (4.35) is: $w'' + [w_o(\xi) - \omega^2]w = 0$. Let w_-, w_+ be the solutions used in the construction of the Green's function G for this equation (see (4.33)). We require more precise estimates for these solutions than the asymptotic estimate in (4.32). To this end, set $\zeta = w'/w$, and form the Riccati equation $\zeta' = \omega^2 - w_o(\xi) - \zeta^2$. We shall estimate w_-; the estimate for w_+ is similar. Let $\zeta = \omega + a(\xi)$, so that $a' = -2\omega a - a^2 - w_o(\xi)$. We augment this with an additional variable $\tau(\xi)$ and form an autonomous system

$$\begin{aligned} a' &= -2\omega a - a^2 + W_o(\tau) \\ \tau' &= \kappa\tau(1-\tau), \end{aligned}$$

where $W_o(\tau) = w_o(\xi(\tau))$ and $\tau = \tau(\xi)$ is the solution of the second equation in the above satisfying $\tau(1/2) = 0$. Note that $\tau(-\infty) = 0$. The exponential decay rate of $w_o(\xi)$ is $\sqrt{b}$. Hence, if $\kappa < \sqrt{b}$ then $W_o(\tau) = \mathbf{O}(\tau^{\sqrt{b}/\kappa})$ near $\tau = 0$. Thus $W_o(\tau)$ is smooth at $\tau = 0$, and $W_o(0) = W_o'(0) = 0$. Furthermore, since $\omega = \sqrt{b(1+\lambda)}$ and $\lambda \in \Lambda$, Re $\omega > K_1|\lambda|^{1/2} > 0$ for some $K_1 > 0$ which is uniformly bounded away from zero for $\lambda \in \Lambda$. The rest point $(a,\tau) = (0,0)$ is therefore a saddle, and we can apply the Unstable Manifold Theorem to the above system to obtain a solution $(a(\xi), \tau(\xi))$ that decays to $(0,0)$ at $\xi = -\infty$. The solution $a(\xi)$ therefore satisfies $|a(\xi)| \le Le^{\kappa\xi}$ for some $L > 0$ as $\xi \to -\infty$.

By Lemma 3.1 we have that $w_o(\xi) \le K$ for some $K > 0$. We then have that the positive function $g(\xi) = [a(\xi)\bar{a}(\xi)]^{1/2}$ satisfies the differential inequality

$$g' \le K_1|\lambda|^{1/2}\left[-g + \frac{g^2}{K_1|\lambda|^{1/2}} + \frac{K}{K_1|\lambda|^{1/2}}\right].$$

Suppose that $1/(K_1|\lambda|^{1/2}) < 1/2$ and that $(K/(K_1|\lambda|^{1/2}) < 1/2$. It then follows that $g' < 0$ if $g = 1$, so that $|a| \le 1$ is positively invariant. Since $a(-\infty) = 0$, it follows that $|a(\xi)| < 1$ for all ξ. The above conditions of λ are equivalent to $|\lambda| \ge H$ for some $\mathbf{O}(1)$ constant $H > 0$. Clearly, $a(\xi)$ must then also have exponential decay at $+\infty$ as well, since the rest point $(0,1)$ is an attractor. The generic (slowest) rate of decay at $+\infty$ is $e^{-\kappa\xi}$. It therefore follows that $|a(\xi)| \le e^{-\kappa|\xi|}$ for all $\xi \in \mathbb{R}$

The solution w_- then satisfies $w_-' = (\omega + a(\xi))w_-$ with $|a(\xi)| \le e^{-\kappa|\xi|}$. In a similar manner, we also obtain that w_+ satisfies $w_+' = (-\omega + b(\xi))w_+$ with $|b(\xi)| \le e^{-\kappa|\xi|}$. The solutions of these equations are then

$$\begin{aligned} w_-(\xi,\lambda) &= e^{\omega\xi + A(\xi)} w_-(0,\lambda), \quad A(\xi) = \int_0^\xi a(r)\,dr \\ w_+(\xi,\lambda) &= e^{-\omega\xi + B(\xi)} w_+(0,\lambda), \quad B(\xi) = \int_0^\xi b(r)\,dr. \end{aligned}$$

Setting $w_\pm(0,\lambda)=1$, it follows that $w'_-(0,\lambda) = \omega + a(0)$ and that $w_+(0,\lambda) = -\omega + b(0)$. Thus $W(\lambda) = -2\sqrt{b(1+\lambda)} + b(0) - a(0)$ with $|a(0)|, |b(0)| \leq 1$. Thus $|1/W(\lambda)| \leq K|\lambda|^{-1/2}$ for $|\lambda| > H$ and $\lambda \in \Lambda$.

We finally note that from the above estimates for $a(\xi), b(\xi)$ we have that $|e^{A(\xi)}|, |e^{B(\xi)}| \leq e^{\kappa^{-1}} < K$ for all ξ. Using these estimates for $w_\pm$ and $W(\lambda)$ in the Green's function representation of w (4.34), we obtain the pointwise estimate

$$\begin{aligned}|w(\xi)| &= \int_{-\infty}^{\infty} G(\xi,y)F(\xi)\\ &\leq \left[\frac{e^{\omega\xi+A(\xi)}}{W(\lambda)}\int_{\xi}^{\infty} e^{-\omega y+B(y)}\,dy + \frac{e^{-\omega\xi+B(\xi)}}{W(\lambda)}\int_{-\infty}^{\xi} e^{\omega y+A(y)}\,dy\right]||F||_\infty\\ &\leq K|\lambda|^{-1}M,\end{aligned}$$

since $\operatorname{Re}\omega \geq K_1|\lambda^{1/2}$. Since $w = v_*$ on $|\xi| \leq 2/\delta^s$, the lemma follows. □

We can now estimate the slow components of y_2 over the transition in the remaining parameter range, $\lambda \in \Lambda_h$ with $|\lambda| \geq H$. To this end, introduce a new parameter $\ell = \delta^\beta\sqrt{\lambda}$, so that the parameter range to be treated is $\delta^\beta\sqrt{H} \leq |\ell| < h\delta^{-r}$. The slow eigenvectors $e_{2,3}(\lambda,\delta)$ can be expressed more simply as

$$e_i(\ell,\delta^2) = (1, \pm\ell + \mathbf{O}(\delta^2), 0, 0,)^t,$$

for $i = 2,3$. The $\mathbf{O}(\delta^2)$ term is of (much) higher order, and it will therefore be suppressed. It is convenient to rescale these eigenvectors by setting

$$\tilde{e}_i(\ell) = (1+|\ell|)^{-1}e_i(\ell).$$

Thus the components of the $\tilde{e}_{2,3}$ are uniformly bounded for all ℓ; in fact, they are unit vectors in the ℓ_1 metric. In the following lemma, we assume that y_2 is expressed in terms of the $\tilde{e}_i(\ell)$'s at $\xi = \pm 2/\delta^s$. It then follows from Lemma 4.5, and the way we have scaled $\tilde{e}_{2,3}$ that

$$\begin{aligned}(4.36)\qquad y_2(-2/\delta^s,\lambda,\delta) &= \tilde{e}_2(\ell) + \mathbf{O}(e^{-\tilde{R}/\delta^\sigma})\\ y_2(+2/\delta^s,\lambda,\delta) &= \tilde{R}_2(\lambda,\delta)\tilde{e}_2(\ell) + \tilde{R}_3(\lambda,\delta)\tilde{e}_3(\ell) + y_{2f}(\lambda,\delta),\end{aligned}$$

where $y_{2f}(\lambda,\delta)$ is the projection of $y_2(2/\delta^s,\lambda,\delta)$ on the fast subspace $\Phi(\lambda,\delta)$, $\ell = \ell(\lambda,\delta) = \delta^\beta\sqrt{\lambda}$, and the constant in the $\mathbf{O}$ term at $-2/\delta^s$ is uniform for $\lambda \in \Lambda$.

THEOREM 4.13. *Suppose that $\lambda \in \Lambda\setminus\Lambda_H$ where Λ_H is as in (3.17), and where $H > 0$ is as in Lemma 3.2 and Theorem 4.11. There exists a (possibly larger) $\mathbf{O}(1)$ value of $H > 0$ and $\delta_0 > 0$ such that $\lambda \in \Lambda\setminus\Lambda_H$ is not an eigenvalue of the perturbed wave $(u_o(\xi,\delta), v_o(\xi,\delta))$ for all $\delta < \delta_0$.*

PROOF. Define the annular region $\Lambda_{H,h} = \lambda_h \cap \{|\lambda| \geq H\}$. We first determine an $\mathbf{O}(1)$ constant $h > 0$ and $\delta_o > 0$ such that the theorem holds for $\lambda \in \Lambda_{H,h}$ for $\delta < \delta_o$. With h now fixed, we then apply Lemma 4.3 to determine $\delta_o > 0$ such that there are no eigenvalues in the complementary region $\Lambda\setminus\Lambda_h$ where δ_o is decreased, if necessary, according to the further restriction imposed by Lemma 4.3.

In the following proof, $K > 0$ will denote a generic $\mathbf{O}(1)$ constant which is bounded from above and bounded away from zero for $\lambda \in \Lambda_H$ and for $\delta < \delta_0$.

We therefore suppose that $H \leq |\lambda| \leq h^2\delta^{2(s-1)}$, or equivalently, $\delta^\beta\sqrt{H} < |\ell| < h\delta^{-r}$. We prove that for some $K > 0$, $|y_{2f}(\lambda,\delta)| \leq Ke^{-\tilde{R}/\delta^\sigma}$ and $|\tilde{R}_2(\lambda,\delta)| \geq 1/2$ for sufficiently small $\delta > 0$.

Clearly, the conclusion of Lemma 4.6 still remains valid if the interval of the fast field is enlarged to $|\xi| \le 1 + 2/\delta^s$. Let $y_*(\xi, \delta, \lambda) \in \hat{y}_2(\xi, \lambda, \delta)$ be scaled as in Lemma 4.6, so that $|y_*| \le M$ for $|\xi| \le 1 + 2/\delta^s$ and all $\lambda \in \Lambda_h$. Suppose that the maximum in (4.16) occurs at ξ_0, and set $(u_{*0}, p_{*0}) = (u_*, p_*)$ at $\xi = \xi_0$. We then have that $|u_{*0}| + |p_{*0}| = 1$. Since $y_*, y_2 \in \hat{y}_2$ both solve (2.4) there exists $\gamma \ne 0$ such that $y_* = \gamma y_2$. By (4.36), $|u| + |p| = 1 + \mathbf{O}(e^{-\hat{R}/\delta^\sigma})$ at $\xi = -2/\delta^s$, and since $|u_*| + |p_*| \le 1$ over the fast field, it follows that $|\gamma| \le 1$. It then follows from (4.36) that

$$y_*(-2/\delta, \lambda, \delta) = \gamma \tilde{e}_2(\ell) + \mathbf{O}(e^{-\hat{R}/\delta^\sigma})$$

uniformly for $\lambda \in \Lambda_h$, since $|\gamma| \le 1$.

Set $(u_{*-}, p_{*-}) = (u_*, p_*)$ at $\xi = -2/\delta^s$. We then have that

$$(u_{*-}, p_{*-}) = \frac{\gamma}{1 + |\ell|}(1, \ell).$$

Let $|\ \ |_\infty$ denote the maximum of a function over the fast field $|\xi| \le 1 + 2/\delta^s$. Since $|u_*|, |p_*| \le 1$ over the fast field, it follows as in Lemma 4.6 that

$$\begin{aligned} (4.37)\quad |u_* - u_{*-}| &\le 4\delta^r |p_*|_\infty \\ |p_* - p_{*-}| &\le 4\delta^r \ell^2 |u_*|_\infty + 4a\delta^{3-s} + \delta^{1-\beta} \int_{-2/\delta^s}^{2/\delta^s} [v_0^2 |u_*| + 2u_0 v_0 |v_*|]\, d\xi \end{aligned}$$

for all $|\xi| \le 1 + 2/\delta^s$. Note that since u_*, v_* are uniformly bounded over the fast field, the integral term is at worst of order $\delta^{1-\beta-s} = \delta^r$.

Since the maximum in (4.16) occurs at ξ_0, we have that either $|u_{*0}| \ge 1/2$ or $|p_{*0}| \ge 1/2$. Assume first that $|u_{*0}| \ge 1/2$, and evaluate the first equation in (4.37) at $\xi = \xi_0$. Since $|p_*|_\infty \le 1$, it follows that $|u_{*-}| \ge 1/4$ if $4\delta^r < 1/4$. Since $|u_{*-}| + |p_{*-}| = \gamma$ we then have that $1/4 \le |\gamma| \le 1$. It also follows from these inequalities and (4.36) that

$$1 \ge |u_{*-}| = |\gamma|(1 + |\ell|)^{-1} \ge 1/[4(1 + |\ell|)],$$

whence $|\ell| \le 3$. Thus the first component of $\tilde{e}_{2,3}(\ell)$ is in $(1/4, 1)$, while the other components are uniformly bounded from above. It then follows from Lemma 4.4 that $|y_{2f}(2/\delta^s, \lambda, \delta)| \le Ke^{-R/\delta^\sigma})$, where K depends only on the metric ρ.

Furthermore, the lower bound for $|\gamma|$ together with the scaling for $|y_*|_\infty < M$ imply that $|y_2|_\infty \le 4M$. Multiplication of (4.37) by γ^{-1} together with the bound for $|y_2|_\infty$ and $|\ell| \le 3$ then yields the estimate

$$\begin{aligned} u(\xi, \lambda, \delta) &= \frac{1}{1 + |\ell|} + \mathbf{O}(\delta^r) \\ p(\xi, \lambda, \delta) &= \frac{\ell}{1 + |\ell|} + \mathbf{O}(\delta^r |\ell|^2) + \mathbf{O}(\delta^{3-s}) + \delta^{1-\beta} \int_{-2/\delta^s}^{\xi} [v_o^2 u + 2u_o v_o v]\, d\xi, \end{aligned}$$

over the fast field. It also follows from the upper bound $|\ell| \le 3$ that Re $(1 + |\ell|)u = 1 + \mathbf{O}(\delta^r) \ge 1/2$ and that $(1 + |\ell|)|v| \le 16$ over the fast field. It then follows from

(4.36) and the above that

$$\begin{aligned}
(\tilde{R}_2+\tilde{R}_3-1) &= \mathbf{O}(\delta^r) \\
(\tilde{R}_2-\tilde{R}_3-1) &= \mathbf{O}(\delta^r|\ell|)+\mathbf{O}(\delta^{3-s}|\ell|^{-1}) \\
&\quad +\frac{(1+|\ell|)\delta^{1-\beta}}{\ell}\int_{-2/\delta^s}^{2/\delta^s}[v_o^2u+2u_ov_ov]\,d\xi \\
&= \mathbf{O}(\delta^r)+\frac{\delta^{1-2\beta}}{\sqrt{\lambda}}\int_{-2/\delta^s}^{2/\delta^s}[v_o^2(1+\mathbf{O}(\delta^r)+2u_ov_o(1+|\ell|)v]\,d\xi,
\end{aligned}$$

where we have used the relation $\ell=\delta^\beta\sqrt{\lambda}$ with $|\lambda|\geq H$. We therefore obtain the expression

$$(4.38)\qquad \tilde{R}_2=1+\mathbf{O}(\delta^r)+\frac{\delta^{1-2\beta}}{2\sqrt{\lambda}}\int_{-2/\delta^s}^{2/\delta^s}[v_o^2(1+\mathbf{O}(\delta^r))+2u_ov_o(1+|\ell|)v]\,d\xi.$$

We now estimate the real part of the integral term in (4.38). Note first that the argument of $\lambda^{-1/2}$ is an angle between $\pm(\pi/4-\mathbf{O}(H^{-1}))$ for $\lambda\in\Lambda_{H,h}$. Thus the coefficient of the integral is a complex number of the form $P(a+ib)$ where $P>0$, $a^2+b^2=1$, and $a>1/\sqrt{2}-\mathbf{O}(H^{-1})$. Also, it follows from the above that $(1+|\ell|)u=1+\mathbf{O}(\delta^r)$ and that $(1+|\ell|)p=\ell+\mathbf{O}(\delta^r)$ uniformly over the fast field. Since $|\ell|\leq 3$, it then follows from Lemma 4.6 that there is a constant $M>0$ such that $|y_2|_\infty\leq M$, where $M>0$ is independent of $\delta>0$ and λ. Thus, by Lemma 4.12, we have that $|v|_\infty\leq KMH^{-1}$ for $\lambda\in\Lambda_{H,h}$. It therefore follows that for sufficiently large but $\mathbf{O}(1)$ $H>0$, the integral term in (4.38) has positive real part. Finally, note that the first term is $1+\mathbf{O}(\delta^r)$. Thus, if $|u_{*0}|\geq 1/2$, we have established that Re $\tilde{R}_2\geq 1/2$ for all $\delta<\delta_o$, and $\lambda\in\Lambda_{H,h}$ for some uniform $\delta_o>0$ and large but $\mathbf{O}(1)$ $H>0$

Next assume that $|p_{*0}|\geq 1/2$ and let $y_*=\gamma y_2$. It again follows from the scaling (4.16) of y_* and by the condition (4.36) for y_2 at $-2/\delta^s$ that $|\gamma|\leq 1$. We then have that $|u_{*-}|=|\gamma u(-2/\delta^s,\lambda,\delta)|\leq(1+|\ell|)^{-1}$. Since $|p_*|_\infty\leq 1$, it then follows from the first equation in (4.37) that

$$|u_*(\xi,\lambda,\delta)|\leq(1+|\ell|)^{-1}+\mathbf{O}(\delta^r),$$

uniformly over the fast field. Using this estimate in the second equation in (4.37) together with the condition that $\delta^r|\ell|\leq h$ we then obtain

$$\begin{aligned}
|p_*(\xi,\lambda,\delta)-p_{*-}| &\leq 4\delta^r|\ell|^2\left[(1+|\ell|)^{-1}+\mathbf{O}(\delta^r)\right]+\mathbf{O}(\delta^{1-\beta}) \\
&\leq 4(h^2+h)+\mathbf{O}(\delta^{1-\beta}),
\end{aligned}$$

since we are assuming that $\delta^r|\ell|\leq h$. We now use the second alternative that $|p_{*0}|\geq 1/2$ at $\xi=\xi_0$. The previous estimate evaluated at $\xi=\xi_0$ then yields $|p_{*-}|\geq 1/2+\mathbf{O}(h)+\mathbf{O}(\delta^r)$. Hence there exists an $\mathbf{O}(1)$ value of $h>0$ such that $|p_{*-}|\geq 1/4$. Recalling that $|\gamma|\leq 1$, we then have that

$$1/4\leq|p_{*-}|=|\gamma||\ell|(1+|\ell|)^{-1}\leq|\ell|(1+|\ell|)^{-1}.$$

We therefore find that now $|\ell|\geq 1/3$, so that $|\ell|$ is uniformly bounded away from zero. The previous inequality involving γ then implies that $1/4\leq|\gamma|\leq 1$ for all $|\ell|\geq 1/3$. Hence γ is uniformly bounded from below. We again conclude that $y_2=\gamma^{-1}y_*$ is uniformly bounded over the fast field.

It now follows as in the derivation of (4.38) that we obtain an expression of the form

$$\tilde{R}_2 = 1 + \mathbf{O}(h) + \frac{(1+|\ell|)\delta^{1-\beta}}{2\ell} \int_{-2/\delta^s}^{2/\delta^s} [v_o^2 u + 2u_o v_o v]\, d\xi.$$

However, now that $|\ell|$ is bounded away from zero, the integral term is uniformly of order $\mathbf{O}(\delta^{1-\beta})$. Hence $|\tilde{R}_2| \geq 1 + \mathbf{O}(h) + \mathbf{O}(\delta^{1-\beta}) \geq 1/2$, for sufficiently small δ and h.

The proof of the Theorem can now be completed for all $\lambda \in \Lambda$. First determine $\mathbf{O}(1)$ values of $\delta_0, h > 0$, (small) and $H > 0$ (large) as above so that Re $\tilde{R}_2 \geq 1/2$ for all $\lambda \in \Lambda_{H,h}$ and $\delta < \delta_0$. It then follows from this, together with Proposition 2.8 and Corollary 3.8 that the bundle $\mathcal{E}(K,\delta)$ can be formed for all $\delta < \delta_o$, and that it can be decomposed into a sum of fast and slow subbundles as in (2.19). By Corollary 3.9, the index of the fast summand is zero. Furthermore, since $\tilde{R}_2$ is uniformly bounded away from zero along $K \cup K^{int}$, the slow subbundle is trivial and has zero index as well, so that by (2.19), $c_1(\mathcal{E}(K,\delta)) = 0$. Hence there are no eigenvalues in the region $\Lambda_{H,h}$.

With $h > 0$ now fixed as above, we then apply Lemma 4.3 to determine a (possibly smaller) choice of $\delta_o > 0$ such that there are no eigenvalues of the wave in $\Lambda \setminus \Lambda_h$. □

4.4. The Construction of $\mathcal{E}_s(K,\delta)$

By Theorem 4.13, unstable eigenvalues of the perturbed wave are confined to the uniformly bounded region Λ_H in (3.17). We next use the slow tracking theorems of the previous sections to construct certain slow subbundles associated to certain simple closed contours $K \subset \Lambda_H$. Let the slow solution $y_2(\xi, \tilde{\lambda}, \delta)$ be defined as in (4.17) in Lemma 4.7. The normalizing condition used in the bundle construction in Chapter 2, section 3 is different in that it requires that

$$e^{-\mu_2 \xi} y_2(\xi, \tilde{\lambda}, \delta) \to e_2(\tilde{\lambda}, \delta) \text{ as } \xi \to -\infty. \tag{4.39}$$

LEMMA 4.14. *There exists an analytic function* $a_2(\tilde{\lambda}, \delta) = \mathbf{O}(e^{-\tilde{R}/\delta^\sigma})$ *and* $\tilde{y}(\xi, \tilde{\lambda}, \delta) \in$ *span*$\{e_1, e_3, e_4\}$ *such that* $y_*(\xi, \tilde{\lambda}, \delta) = \mathbf{O}(e^{-\tilde{R}/\delta^\sigma})$ *and*

$$e^{-\mu_2 \xi} y_2(\xi, \tilde{\lambda}, \delta) = [1 + a_2(\tilde{\lambda}, \delta)] e_2(\tilde{\lambda}, \delta) + \tilde{y}(\xi, \tilde{\lambda}, \delta).$$

This estimate holds uniformly for $\xi < -2/\delta^s$ *and for* $\tilde{\lambda}$ *in compact subsets of* Λ.

PROOF. Since $y_2 \in \hat{y}_2$ and $\hat{y}_2 \to \hat{e}_2$ as $\xi \to -\infty$, Lemma 2.5 of **[14]** implies that $e^{-\mu_2 \xi} y_2(\xi, \tilde{\lambda}, \delta)$ is uniformly bounded on the half line $\xi < -2/\delta^s$, and that it in fact has a limit $a_2(\tilde{\lambda}, \delta) e_2(\tilde{\lambda}, \delta)$ that is a scalar multiple of e_2. Express $y_2 = \Sigma_{i=1}^4 (L_i + a_i(\xi, \tilde{\lambda}, \delta)) w_i(\xi, \tilde{\lambda})$ as in the proof of Lemma 4.5, where $w_i = e_i^{\mu_i(\xi + 2/\delta^s)}(\tilde{\lambda}, \delta)$ and each $a_i = 0$ at $\xi = -2/\delta^s$. From the normalization (4.17) we have that $L_2 = 1$ and $L_i = \mathbf{O}(e^{-\tilde{R}/\delta^\sigma})$ for $i = 1, 3, 4$. It then follows as in the proof of Lemma 4.5 that $|a_2(\xi)|, |a_3(\xi)| = \mathbf{O}(e^{-\tilde{R}/\delta^s})$ for all $\xi < -2/\delta^s$. Thus

$$e^{-\mu_2 \xi} y_2(\xi, \tilde{\lambda}, \delta) = [1 + \mathbf{O}(e^{-\tilde{R}/\delta^s})] e_2(\tilde{\lambda}, \delta) + \mathbf{O}(e^{-\tilde{R}/\delta^\sigma}) e_3(\tilde{\lambda}, \delta) + P_f(\xi)$$

where $P_f(\xi)$ is the projection of $e^{-\mu_2 \xi} y_2$ on the fast subspace. Since the coefficient of e_2 is uniformly bounded away from zero, it follows from Lemma 4.4 that $|P_f(\xi)| = \mathbf{O}(e^{-R/\delta^\sigma})$ for all $\xi \leq -2/\delta^s$. The statement of the lemma then holds with $\tilde{y}(\xi, \tilde{\lambda}, \delta) = (L_3 + a_3) e_3 + P_f(\xi)$. □

It now follows that the solution $\bar{y}_2(\xi,\tilde{\lambda},\delta) = (1+a_2(\tilde{\lambda},\delta))^{-1}y_2(\xi,\tilde{\lambda},\delta)$ satisfies (4.39). Since a_2 is exponentially small, we see that the transmission coefficient $\bar{R}_2(\tilde{\lambda},\delta)$ for $\bar{y}_2$ at $\xi = 2/\delta^s$ differs from $R_2(\tilde{\lambda},\delta)$ by an exponentially small term. We can therefore drop the "bar" and assume that y_2 satisfies (4.39) instead of (4.17), and that the additional exponentially small error is absorbed into the algebraic error terms in (4.30).

Using the notation of Chapter 2, section 3, set $\alpha(\tau,\tilde{\lambda},\delta) = e^{-\mu_2\xi(\tau)}y_2(\xi(\tau),\tilde{\lambda},\delta)$ for $-1 < \tau < \tau_+$ and for $\tilde{\lambda} \in K$, and set $\alpha(-1,\tilde{\lambda},\delta)$, $\beta(\tilde{\lambda},\delta) = e_2(\tilde{\lambda},\delta)$ for $\tilde{\lambda} \in K \cup K^{int}$. It then follows that a slow subbundle $\mathcal{E}_s(\tilde{\lambda},\delta)$ can be constructed as in Chapter 2, section 3 with clutching function $R_2(\tilde{\lambda},\delta)$. We summarize this in the following corollary.

COROLLARY 4.15. *Suppose that $K \subset \Lambda$ is a simple closed curve and that $R_2(\tilde{\lambda},\delta)$ is uniformly bounded away from zero for all $\tilde{\lambda} \in K$ and for all sufficiently small δ, where R_2 is the analytic expression given by (4.30), (4.29). Then the bundle $\mathcal{E}(K,\delta)$ is defined for all sufficiently small $\delta > 0$, and it contains a one dimensional subbundle $\mathcal{E}_s(K,\delta)$ with clutching function $R_2(\tilde{\lambda},\delta)$ such that $\mathcal{E}(K,\delta) = \mathcal{E}_f(K,\delta) \oplus \mathcal{E}_s(K,\delta)$, where $\mathcal{E}_f(K,\delta)$ is as defined in Chapter 3, section 4.*

REMARK 4.16. When $K \subset S_1$, R_2 is determined by taking the plus sign in (4.29). In this case, the asymptotic flow for the projectivized system at $+\infty$ on the slow subspace $\hat{\Sigma}$ forms an attractor-repeller pair $(\hat{e}_3,1) \to (\hat{e}_2,1)$. Since $(\hat{y}_2(\xi,\tilde{\lambda},\delta),\tau(\xi)) \to (\hat{e}_2(\tilde{\lambda},\delta),-1)$ as $\xi \to -\infty$, this solution generically tends to $(\hat{e}_2(\tilde{\lambda},\delta),1)$ at $+\infty$. It is not difficult to show that this situation occurs using the slow tracking machinery, whenever $R_2(\lambda,\delta)$ in (4.30) is bounded away from zero. However, when K intersects the domain with negative spectral gap, the asymptotic flow restricted to $\hat{\Sigma}$ is now the attractor-repeller pair $\hat{e}_2 \to \hat{e}_3$. In particular, this occurs along the second sheet of $\tilde{K} = \partial\tilde{N}$. Along this sheet, it is not difficult using Gap Lemma arguments similar to those of the proof of Lemma 4.5 to show that $(\hat{y}_2,\tau) \to (\hat{e}_3,1)$ at $+\infty$ whenever $R_3(\tilde{\lambda},\delta)$ in (4.30) is bounded away from zero. Hence $\hat{y}_2(+\infty,\tilde{\lambda},\delta)$ is discontinuous across the branch cut. In fact, when $\tilde{\lambda}$ lies on the branch cut, $\hat{y}_2(\xi,\tilde{\lambda},\delta)$ is asymptotically periodic in ξ. This is the reason that the bundle construction in **[1]** fails whenever K intersects the region with negative spectral gap.

CHAPTER 5

Calculation of the Stability Index

We now complete the proof of Theorem 1.3 by calculating the stability index $c_1(\mathcal{E}(K,\delta))$ for various curves K in the spectral plane in each of the three regimes, $0 \le \beta < 1/2$, $\beta = 1/2$, and $1/2 < \beta < 1$. In order to form the bundle and calculate its index, we first need to to relate the clutching function R_2 (see Corollary 4.15) for the slow summand to the solutions of the NLEP, (1.13).

The integral operator $I_*(v_*, \tilde{\lambda}, \delta)$ in (4.29) can be explicitly determined from the hypergeometric function expression for v_* as a meromorphic function of $\tilde{\lambda}$. This function is closely related to the integral operator c in (1.11) arising in the solution of the NLEP. Note that c can be analytically continued to the Riemann surface by defining

$$c(v_*;\tilde{\lambda},\delta) = \frac{-\delta^{1-2\beta}\bar{c}^3}{3b^{3/2}\delta^{1-2\beta} \pm \bar{c}^2\sqrt{b(\lambda-\lambda_*(\delta))} + \mathbf{O}(\delta^{2-\beta})} \int_{-\infty}^{+\infty} S(\xi)v_*(\xi,\tilde{\lambda})\, d\xi \tag{5.1}$$

where $\tilde{\lambda} = (\lambda, \pm\sqrt{\lambda - \lambda_*(\delta)})$ in the local coordinate λ, and where the plus/minus sign is determined in the usual way. For $\tilde{\lambda} \in \Lambda$, the error term in the denominator results from replacing $\sqrt{b\lambda}$ in c by $\sqrt{b(\lambda-\lambda_*(\delta))}$ and using that $\lambda_*(\delta) = \mathbf{O}(\delta^{2-\beta})$.

An explicit calculation shows that

$$\int_{-\infty}^{\infty} S(\xi)^2\, d\xi = \frac{6b\sqrt{b}}{\bar{c}^2}, \tag{5.2}$$

(see (4.12) of [6]). Now let

$$R_{2*}(\tilde{\lambda},\delta) = 1 + I_*(v_*,\tilde{\lambda},\delta), \tag{5.3}$$

so that $|R_2 - R_{2*}| = \mathbf{O}(\delta^{\tilde{m}(\delta)})$ with $\tilde{m}(\delta) > 1 - 2\beta$, by Theorem 4.11. Thus R_{2*} differs from R_2 by terms that are of higher order. Furthermore, R_{2*} can be determined explicitly from the hypergeometric solutions of (1.13). In particular, the following lemma relates R_{2*} to the meromorphic function $f(\tilde{\lambda},\delta) = 1 - c(v_*,\tilde{\lambda},\delta)$, see 1.15), whose roots determine the eigenvalues of the NLEP (1.13).

LEMMA 5.1. *Let $\tilde{\lambda} = (\lambda, \pm\sqrt{(\lambda-\lambda_*(\delta))})$. $R_{2*}(\tilde{\lambda},\delta)$ be as in (5.3) with I_* given by (4.29), and let $f(\tilde{\lambda},\delta) = 1 - c(v_*(\tilde{\lambda},\delta),\tilde{\lambda},\delta)$. Then*

$$\begin{aligned} R_{2*}(\tilde{\lambda},\delta) &= \frac{1}{\pm\bar{c}^2\sqrt{\lambda-\lambda_*(\delta)}}\left[3b\delta^{1-2\beta} \pm \bar{c}^2\sqrt{\lambda-\lambda_*(\delta)} + \mathbf{O}(\delta^{2-\beta})\right] f(\tilde{\lambda},\delta) \\ &\quad + \mathbf{O}(\delta^{2-\beta}) \end{aligned} \tag{5.4}$$

for all $\tilde{\lambda}$ in compact subsets of Λ.

PROOF. The lemma is easily verified from (5.1), (5.2), (5.3), and (4.29); we omit the algebraic details. □

5.1. The Strongly Unstable Regime: $0 \leq \beta < 1/2$

In this section, we prove part (i) of Theorem 1.3. Since we only need to show the presence of (at least one) positive eigenvalue, the analysis near $\lambda = 0$ is not relevant to this discussion. We therefore take the "plus" sign in all formulae involving $\sqrt{\lambda - \lambda_*(\delta))}$ in this discussion.

Let $K \subset \Lambda \cap S_1$ be a simple closed curve contained in the right half plane $\mathrm{Re}\,\lambda > 0$ of the standard spectral plane and which encloses the positive eigenvalue $\lambda_1 = 5/4$ of the Fisher wave $S(\xi)$ in its interior and which is disjoint from the neighborhood $\tilde{N}$ of the origin. Here and below the curve K remains fixed as $\delta \to 0$. We also require that K be disjoint from the eigenvalues of the NLEP (1.13). It is shown in **[6]** that the NLEP has exactly two unstable eigenvalues $\lambda_{iN}(b)$, $i = 0, 1$ in this regime (see Chapter 1, section 3); $\lambda_{0N}(b)$ is asymptotically close to the origin, while $\lambda_{1N}(b)$ is asymptotically close to λ_1. The curve K therefore contains the eigenvalue $\lambda_{1N}(b)$ in its interior for all small $\delta > 0$; the other eigenvalue $\lambda_{0N}(b)$ is exterior to K.

Since $K \subset \Lambda$, the results of Chapter 3 are valid for such K, and so $\mathcal{E}_f(K, \delta)$ can be formed, and by Corollary 3.9, we have that $c_1(\mathcal{E}_f(K, \delta)) = 1$. Furthermore, since K is uniformly bounded away from the eigenvalue λ_1 of the Fisher wave, the solution $v_*(\xi, \lambda)$ of the differential equation in (1.13) exists and decays at $\pm\infty$. The integral operator I_* is therefore uniformly $\mathbf{O}(\delta^{1-2\beta})$ for $\lambda \in K$. The transmission coefficient $R_2(\lambda, \delta)$, for y_2 is asymptotically close to unity along K, since $\beta < 1/2$. It therefore follows that Corollary 4.15 is applicable for such contours, and we can therefore construct the slow summand $\mathcal{E}_s(K, \delta)$ and the full bundle $\mathcal{E}_s(K, \delta)$ with clutching function $R_2(\lambda, \delta)$.

We now apply the sum formula (2.19) together with the fast index calculation (Corollary 3.9) to obtain

$$c_1(\mathcal{E}(K, \delta)) = 1 + c_1(\mathcal{E}_s(K, \delta)).$$

It only remains to calculate the index of the slow summand. In the strongly unstable regime $\beta < 1/2$ this follows immediately from the observation that the clutching function $R_2(\lambda, \delta)$ for the slow summand is uniformly close to unity along K; consequently $c_1(\mathcal{E}_s(K, \delta)) = 0$, so that $c_1(\mathcal{E}(K, \delta)) = 1$ for all sufficiently small $\delta > 0$. Hence, K encloses at exactly one unstable eigenvalue of the perturbed wave, which establishes case (i) of Theorem 1.3.

REMARK 5.2. It is actually the case that the transmission coefficient R_2 has a simple pole and also a simple root interior to K. This follows from the fact that the NLEP has a simple eigenvalue that is asymptotically close to λ_1 in this regime (see Chapter 1, section 2). Hence, the net contribution of the slow subbundle to the full index is zero.

5.2. The Other Regimes: $1/2 \leq \beta < 1$

Since we now have the possibility of stable waves, it is necessary to give a precise count of eigenvalues near the origin.

LEMMA 5.3. *The following results hold for all $\delta < \delta_o$, for some $\delta_o > 0$:*

(i) Suppose that $1/2 \leq \beta < 1$. There exists a small, fixed $\mathbf{O}(1)$ neighborhood $\tilde{N}$ of the branch point $\lambda_*(\delta)$ on $\mathcal{R}$ such that $R_2(\tilde{\lambda}, \delta)$ is uniformly bounded away from zero along the contour $\tilde{K} = \partial\tilde{N}$.

(ii) The bundles $\mathcal{E}_s(\tilde{K},\delta)$ and $\mathcal{E}(\tilde{K},\delta)$ can be formed, $c_1(\mathcal{E}_s(\tilde{K},\delta)) = -1$, and $c_1(\mathcal{E}(\tilde{K},\delta)) = 1$.

(iii) Let N be the projection of $\tilde{N}$ on the complex plane, S_1. The only eigenvalue of the perturbed wave in N is the translational eigenvalue at the origin, and $\lambda = 0$ is a simple eigenvalue.

PROOF. (i) By Theorem 4.11, we have that $R_2 = R_{2*} + \mathbf{O}(\delta^{\tilde{m}(\beta)})$ with $\tilde{m}(\beta) > 1 - 2\beta$ and $R_{2*} = 1 + I_*$, where I_* is the operator in (4.29). A straight-forward calculation shows that the solution $v_*(\xi, 0)$ of the differential equation in (1.13) at $\lambda = 0$ is explicitly given by

$$v_*(\xi,0) = -\frac{a}{6b^2}\text{sech}^2(\sqrt{b}\xi/2) = -\frac{\sqrt{a}}{3b^{3/2}}S(\xi).$$

Using the explicit formulae (1.7) for $\bar{c}, S(\xi)$ together with (5.2), we then have that the integral in I_* at $\lambda = 0$ is therefore

$$\int_{-\infty}^{\infty} [S(\xi)^2 + 2\bar{c}S(\xi)v_*(\xi,0)]\, d\xi = -\int_{-\infty}^{\infty} S(\xi)^2\, d\xi = -\frac{2a}{3b^{3/2}}.$$

It then follows that

$$R_2(\tilde{\lambda},\delta) = 1 + \delta^{1-2\beta}\left\{\frac{-a}{\pm 3b^2\sqrt{(\lambda - \lambda_*(\delta))}} + \delta^{2\beta-1}\left[\mathbf{O}(|\lambda|) + \mathbf{O}(\delta^{\tilde{m}(\beta)})\right]\right\}, \tag{5.5}$$

where λ is the projection of $\tilde{\lambda}$ on S_1. Since $\beta \in [1/2, 1)$, $2\beta - 1 + \tilde{m}(\beta) > 0$, and $\tilde{\lambda} = (\lambda, \pm\sqrt{\lambda - \lambda_*(\delta)}) \in \partial\tilde{N}$ is a small but $\mathbf{O}(1)$, it follows that $R_2(\tilde{\lambda},\delta)$ is uniformly bounded away from zero along $\tilde{K} = \partial\tilde{N}$.

(ii) Since $R_2(\tilde{\lambda},\delta)$ is bounded away from zero along $\tilde{K}$, Corollary 4.15 implies that the slow summand $\mathcal{E}_s(\tilde{K},\delta)$ and also the full bundle, $\mathcal{E}(\tilde{K},\delta)$, can be formed for such contours $\tilde{K}$. Thus the sum formula (2.19) holds, where $\mathcal{E}_f(\tilde{K},\delta)$ is the bundle of part (ii) of Corollary 3.9.

We first calculate $c_1(\mathcal{E}_s(\tilde{K},\delta))$. We let $G(\zeta) = \frac{-a}{3b^2\zeta}$, and also we let $g(\tilde{\lambda}) = G(\pm\sqrt{\lambda - \lambda_*(\delta)})$, where $\tilde{\lambda} = (\lambda, \pm\sqrt{\lambda - \lambda_*(\delta)})$ is as in Theorem 2.10, so that $Z = P = 0$ and $\ell = -1$. It then follows from (5.5) and Theorem 2.10 that

$$c_1(\mathcal{E}_s(\tilde{K},\delta)) = W(R_2(\tilde{K},\delta)) = W(g(\tilde{K})) = -1.$$

Combining this result with the sum formula (2.19) and the fact that $c_1(\mathcal{E}_f(\tilde{K},\delta)) = 2$ (see Corollary 3.9), we finally obtain $\mathcal{E}(\tilde{K},\delta) = 1$.

(iii) The proof is completed by combining the calculation in (ii) with Corollary 2.7. In particular, we have that $c_1(\mathcal{E}(\tilde{K},\delta)) = W(D(\tilde{K},\delta)) = 1$, while from Corollary 2.7, this index gives an upper bound for the multiplicity of eigenvalues of the wave in N. Thus $\lambda = 0$ is the unique eigenvalue of the perturbed wave in N, and it is algebraically simple. □

COROLLARY 5.4. *There exists $\delta_o > 0$ such that the multiplicity of eigenvalues of L in the region Λ defined in (3.17) is exactly three for all $\delta < \delta_o$.*

PROOF. By Theorem 4.13, there exists an $\mathbf{O}(1)$ constant $H > 0$ and $\delta_o > 0$ such that L has no eigenvalues in the region $\Lambda \cap \{|\lambda| \geq H\}$ for all $\delta < \delta_o$. Let C_H be the contour used in the numerical winding number argument of Proposition 1.1, see Figure 1.2. It follows from Corollary 3.9 that $\mathcal{E}_f(C_H,\delta)$ exists for $\delta < \delta_o$ and that the fast summand has index $+1$, since C_H encloses λ_1 in its interior. It also follows from Corollary 4.15, Lemma 5.1, and Proposition 1.1 that $\mathcal{E}_s(C_H, \delta)$ and $\mathcal{E}(C_H, \delta)$ can both be formed for $\delta < \delta_o$, since $f(\lambda, \delta)$, and hence $R_2(\lambda, \delta)$, is nonvanishing along C_H. Hence $c_1(\mathcal{E}(C_H, \delta)) = 1 + c_1(\mathcal{E}_s(C_H, \delta))$.

By (i) of Theorem 2.10, we have that $c_1(\mathcal{E}_s(C_H, \delta)) = Z - P$, where Z (resp. P) is the multiplicity of roots (resp. poles) of R_2 interior to C_H. By Proposition 1.1, the NLEP has exactly two roots interior to C_H. By Theorem 4.11 and Lemma 5.1, the same must be true of R_2, so that $Z = 2$.

We also claim that $P = 1$. Clearly, R_2 is analytic near any point λ for which the differential equation in (1.13) is solvable. For points interior to C_H, the only point where this equation is not solvable is at the principal eigenvalue $\lambda_1 = 5/4$ of the Fisher wave. This is because the inhomogeneous term $-S(\xi)^2$ is strictly negative, while the eigenfunction associated to λ_1 is positive. Since the principal eigenvalue of a scalar problem is necessarily simple, it follows that the Wronskian $W(\lambda)$ in the Green's function (4.33) for the linearized Fisher wave has a simple root at λ_1, and hence, the resolvent operator $v_*(\xi, \lambda)$ in (4.34) and also I_* (see (4.29)) have simple poles there. It then follows from Theorem 4.11 and the above that $R_2 = 1 + I_* + \mathbf{O}(\delta^{\tilde{m}(\beta)})$ also has a simple pole in a small neighborhood of λ_1, so that $P = 1$.

It then follows from Theorem 2.10 that $c_1(\mathcal{E}_s(C_H, \delta)) = 1$, so that the stability index $c_1(\mathcal{E}(C_H, \delta)) = 2$. We therefore have that the total multiplicity of eigenvalues of the perturbed wave interior to C_H is two. By the previous lemma, there is exactly one simple eigenvalue in the region N at the origin, so that the total multiplicity of eigenvalues interior to Λ is three. □

The proof of Theorem 1.3 can now be completed for both of the regimes $\beta = 1/2$ and $1/2 < \beta < 1$. In both cases we choose small, $\mathbf{O}(1)$ neighborhoods $N_i(b) \subset \Lambda \backslash \tilde{N}$, $i = 0, 1$ which enclose the two NLEP eigenvalues $\lambda_{iN}(b)$ in their interiors, for $i = 0, 1$. In the critical regime, a minor technical difficulty occurs for b near b_C, where $\lambda_{0N}(b_C) = \lambda_{1N}(b_C)$. Here we use a single neighborhood for b near b_C, and then replace this with the union of two distinct neighborhoods for other values of b.

For $b \neq b_C$ we obtain from our index calculation that $c_1(\mathcal{E}(\partial N_i(b), \delta)) = 1$ for $i = 0, 1$, so that these neighborhoods track the eigenvalues of the perturbed wave for small $\delta > 0$ as b is decreased, as depicted in Figure 1.1. This completes the proof of Theorem 1.3.

CHAPTER 6

Concluding Remarks

The stability index analysis of the homoclinic solutions of the Gray-Scott equations exhibits two new phenomena: the pulse-stabilizing zero-pole cancellation in the NLEP equation and the multivalued Evans function. Neither one of them is special to the Gray-Scott model.

It is shown in [**7**] that a large class of two-dimensional slow/fast reaction diffusion systems on the real line has singular homoclinic solutions with structure and behavior similar to that in the Gray-Scott model. This class includes the classical and generalized Gierer-Meinhardt equations, the Schnakenberg equations, and the P-model in [**23**]. As in Gray-Scott, the pulse solutions correspond to unstable stationary homoclinic solutions of the scalar fast reduction. However, these solutions can be stabilized by the coupling to the slow equation through a nonlocal NLEP mechanism. The Evans function associated to the stability problem can be decomposed into a product. Once again, the zero of the first factor, the unstable eigenvalue of the fast reduced limit, is annihilated by a pole in the second, slow, component.

The stability of the homoclinic patterns in this paper can be seen as a first step towards a fundamental geometrical understanding of more general singular pulse-like patterns in reaction-diffusion systems. In [**8**] it has been shown that there also exist singular spatially periodic and multi-loop homoclinic solutions to the Gray-Scott equations. Similar patterns are constructed for more general systems in [**7**]. The periodic solutions have been observed numerically as stable singular patterns [**6**]. Moreover, these patterns can be interpreted as the endproducts of the transient self-replicating pulse phenomenon [**8, 23**]. In [**7**] the NLEP method has been used to study the stability of these patterns. The analysis is at leading order very similar to that for the homoclinic patterns. The NLEP procedure yields a critical value of b at which a periodic pattern can become stable by a Hopf bifurcation. This value agrees with the numerical observations. However, in the present form, the NLEP method cannot be used to study the spectrum of the linearized stability problem near $\lambda = 0$ because the portion of the spectrum of the wave near the origin will generally consist of a loop of eigenvalues rather than a single simple eigenvalue. This was rigorously established in [**11**] for long wavelength periodic waves by an analysis of the Evans function. Determining the precise location of the critical loop of spectrum is a difficult issue. This was recently accomplished for the periodic solutions of the FitzHugh-Nagumo system see [**9**] near the singular limit. Although the singular limit of the Gray-Scott system is substantially more complicated, a promising line of analysis of the periodic Gray-Scott solutions would be to combine the NLEP approach with the methods in [**11**] and [**9**].

In [**7**] the ideas developed in this paper and in [**6**] have been expanded to study the stability of the multi-pulse homoclinic orbits. As noted in the Introduction, the Gray-Scott equation (1.1) also possesses stationary multiple-pulse solutions, see Theorem 4.1 in [8]. These $N-$pulse solutions consist of N narrow intervals on which the V component exhibits a large-amplitude pulse that is close to the single pulse/Fisher wave, and on which the U component is approximately constant, $3\delta^{3\beta/2}Nb\sqrt{b/a}$. Recently, it has been proven that multi-pulse solutions are unstable for any $N \geq 2$, see [7].

The key mechanism behind the instability of the multi-pulse homoclinic pulses can be understood by the methods developed in this paper: the decomposition of the Evans function into a fast analytic component and a slow component with singularities. The methods of Chapter 3 show that the fast subbundle associated to an N-pulse solution will have an index of N for every $N \geq 2$. However, the methods of Chapter 4 apply to the analysis of the slow subbundle associated to the curve encircling the reduced Fisher eigenvalue at $\lambda = 5/4$. Consequently, this subbundle always has a winding number of -1, so that only one of the zeroes of the fast subsystem can be cancelled by the pole produced by the NLEP.

A similar instability mechanism occurs in certain scalar reaction diffusion equations that have a nonlocal term, see [3]. Also, we remark that the results on the stability of multi-pulse solutions in reaction-diffusion systems presented in [28] have been obtained for a class of multi-circuit homoclinic orbits that differs significantly from the orbits appearing in the Gray-Scott model or in the reaction-diffusion equations studied in [7] (see [7] for more details).

Bibliography

[1] J. Alexander, R. Gardner, and C.K.R.T. Jones. A topological invariant arising in the stability analysis of travelling waves. *J reine angew. Math*, 410:167–212, 1990.

[2] M. F. Atiyah. *K-Theory.* Addison-Wesley Publishing Company, 1989.

[3] A. Bose and G. A. Kriegsmann. Stability of localized structures in non-local reaction-diffusion equations. *Meth. Appl. Anal.*, 5:351–366, 1998.

[4] A. Doelman, W. Eckhaus, and T.J. Kaper. Slowly-modulated two pulse solutions in the 1-d gray-scott model i: asymptotic construction and stability. 2000, to appear *SIAM J. Appl. Math.*.

[5] A. Doelman, W. Eckhaus, and T.J. Kaper. Slowly-modulated two pulse solutions in the gray-scott model ii: geometric theory, bifurcations, and splitting dynamics. 2000, to appear *SIAM J. Appl. Math.*.

[6] A. Doelman, R. A. Gardner, and T. J. Kaper. Stability of singular patterns in the 1-d gray-scott model : A matched asymptotic approach. *Physica D*, 122:1–36, 1998.

[7] A. Doelman, R.A Gardner, and T.J. Kaper. Large stable pulse solutions in reaction-diffusion equations. 2000, to appear *Ind. Univ. Math. J.*.

[8] A. Doelman, T. Kaper, and P. Zegeling. Pattern formation in the 1-d gray-scott model. *Nonlinearity*, 10:523–563, 1997.

[9] E. Eszter. *An Evans function analysis of the stability of periodic travelling wave solutions of the FitzHugh-Nagumo system.* PhD thesis, University of Massachusetts, Amherst, 1999.

[10] N. Fenichel. Geometric singular perturbation theory for ordinary differential equations. *J. Diff. Equations*, 31:53–98, 1979.

[11] R. A. Gardner. Instability of large wavelength periodic wavetrains in reaction-diffusion systems. In P. Bates, S-N Chow, and X. Pan, editors, *Differential Equations and Applications*, pages 82–95. International Press, 1996.

[12] R. A. Gardner and C.K.R.T Jones. Traveling waves of a perturbed diffusion equation arising in a phase field model. *Indiana Univ. Math. J.*, 38:1197–1222, 1989.

[13] R. A. Gardner and C.K.R.T. Jones. Stability of travelling waves of diffusive predator prey systems. *T.A.M.S*, 327:465–524, 1991.

[14] R. A. Gardner and K. Zumbrun. A geometric condition for linearized stability of undercompressive viscous shock waves. *C. P. A. M.*, 51:797–855, 1998.

[15] P. Gray and S. Scott. Autocatalytic reactions in the isothermal, continuous stirred tank reactor: isolas and other forms of multistability. *Chem. Engin. Sci.*, 38:29–43, 1983.

[16] P. Gray and S. Scott. Autocatalytic reations in the isothermal, continuous stirred thank reactor: oscillations and instabilities in sthe system $a + 2b \to 3b, b \to c$. *Chem. Engin. Sci.*, 38:1087–1097, 1984.

[17] P. Gray and S. Scott. Sustained oscillations and other exotic patterns of behavior in isothermal reactions. *J. Phys. Chem.*, 89:22–32, 1985.

[18] J.K. Hale, L.A. Peletier, and W.C. Troy. Exact homoclinic and heteroclinic solutions of the gray-scott model for autocatalysis. to appear: S.I.A.M. J. Appl. Math.

[19] D. Henry. *Geometric Theory of Semilinear Parabolic Equations*, volume 840 of *Springer lecture notes in math.* Springer-Verlag, 1981.

[20] C.K.R.T. Jones. Stability of the traveling wave solution of the fitzhugh-nagumo system. *Transactions Amer. Math. Soc.*, 286:431–469, 1984.

[21] T. Kapitula and B. Sandstede. Stability of bright and dark solitary-wave solutions to perturbed nonlinear schrodinger equations. *Physica D*, 124:58–103, 1998.

[22] C.B Muratov and V.V. Osipov. Spike autosolitons in the gray-scott model. preprint, 1998.

[23] Y. Nishiura and D. Ueyama. A skeleton structure for self-replication dynamics. *Physica D*, 130:73–104, 1999.

[24] J.E. Pearson. Complex patterns in a simple system. *Science*, 261:189–192, 1993.

[25] V. Petrov, S.K. Scott, and K. Showalter. Excitability, wave reflection, and wave splitting in a cubic autocatalysis reaction-diffusion system. *Phil. Trans. Roy. Soc. Lond., Series A*, 347:631–642, 1994.

[26] W. Reynolds, J. Pearson, and S. Ponce-Dawson. Dynamics of self-replicating patterns in reaction-diffusion systems. *Phys. Rev. Letters*, 72:2797–2800, 1994.

[27] J. E. Rubin. *The generation of edge oscillations in an inhomogeneous reaction-diffusion system.* PhD thesis, Brown University, 1996.

[28] B. Sandstede. Stability of multiple-pulse solutions. *Transactions A. M. S.*, 350:429–472, 1998.

Editorial Information

To be published in the *Memoirs*, a paper must be correct, new, nontrivial, and significant. Further, it must be well written and of interest to a substantial number of mathematicians. Piecemeal results, such as an inconclusive step toward an unproved major theorem or a minor variation on a known result, are in general not acceptable for publication. Papers appearing in *Memoirs* are generally longer than those appearing in *Transactions*, which shares the same editorial committee.

As of September 30, 2001, the backlog for this journal was approximately 7 volumes. This estimate is the result of dividing the number of manuscripts for this journal in the Providence office that have not yet gone to the printer on the above date by the average number of monographs per volume over the previous twelve months, reduced by the number of volumes published in four months (the time necessary for preparing a volume for the printer). (There are 6 volumes per year, each containing at least 4 numbers.)

A Consent to Publish and Copyright Agreement is required before a paper will be published in the *Memoirs*. After a paper is accepted for publication, the Providence office will send a Consent to Publish and Copyright Agreement to all authors of the paper. By submitting a paper to the *Memoirs*, authors certify that the results have not been submitted to nor are they under consideration for publication by another journal, conference proceedings, or similar publication.

Information for Authors

Memoirs are printed from camera copy fully prepared by the author. This means that the finished book will look exactly like the copy submitted.

The paper must contain a *descriptive title* and an *abstract* that summarizes the article in language suitable for workers in the general field (algebra, analysis, etc.). The *descriptive title* should be short, but informative; useless or vague phrases such as "some remarks about" or "concerning" should be avoided. The *abstract* should be at least one complete sentence, and at most 300 words. Included with the footnotes to the paper should be the 2000 *Mathematics Subject Classification* representing the primary and secondary subjects of the article. The classifications are accessible from `www.ams.org/msc/`. The list of classifications is also available in print starting with the 1999 annual index of *Mathematical Reviews*. The Mathematics Subject Classification footnote may be followed by a list of *key words and phrases* describing the subject matter of the article and taken from it. Journal abbreviations used in bibliographies are listed in the latest *Mathematical Reviews* annual index. The series abbreviations are also accessible from `www.ams.org/publications/`. To help in preparing and verifying references, the AMS offers MR Lookup, a Reference Tool for Linking, at `www.ams.org/mrlookup/`. When the manuscript is submitted, authors should supply the editor with electronic addresses if available. These will be printed after the postal address at the end of the article.

Electronically prepared manuscripts. The AMS encourages electronically prepared manuscripts, with a strong preference for $\mathcal{A}\mathcal{M}\mathcal{S}$-LaTeX. To this end, the Society has prepared $\mathcal{A}\mathcal{M}\mathcal{S}$-LaTeX author packages for each AMS publication. Author packages include instructions for preparing electronic manuscripts, the *AMS Author Handbook*, samples, and a style file that generates the particular design specifications of that publication series. Though $\mathcal{A}\mathcal{M}\mathcal{S}$-LaTeX is the highly preferred format of TeX, author packages are also available in $\mathcal{A}\mathcal{M}\mathcal{S}$-TeX.

Authors may retrieve an author package from e-MATH starting from `www.ams.org/tex/` or via FTP to `ftp.ams.org` (login as `anonymous`, enter username as password, and type `cd pub/author-info`). The *AMS Author Handbook* and the *Instruction Manual* are available in PDF format following the author packages link from `www.ams.org/tex/`. The author package can be obtained free of charge by sending email to `pub@ams.org` (Internet) or from the Publication Division, American Mathematical Society, P.O. Box 6248, Providence, RI 02940-6248. When requesting an author package, please specify $\mathcal{A}\mathcal{M}\mathcal{S}$-LaTeX or $\mathcal{A}\mathcal{M}\mathcal{S}$-TeX, Macintosh or IBM (3.5) format, and the publication in which your paper will appear. Please be sure to include your complete mailing address.

Sending electronic files. After acceptance, the source file(s) should be sent to the Providence office (this includes any TeX source file, any graphics files, and the DVI or PostScript file).

Before sending the source file, be sure you have proofread your paper carefully. The files you send must be the EXACT files used to generate the proof copy that was accepted for publication. For all publications, authors are required to send a printed copy of their paper, which exactly matches the copy approved for publication, along with any graphics that will appear in the paper.

TeX files may be submitted by email, FTP, or on diskette. The DVI file(s) and PostScript files should be submitted only by FTP or on diskette unless they are encoded properly to submit through email. (DVI files are binary and PostScript files tend to be very large.)

Electronically prepared manuscripts can be sent via email to `pub-submit@ams.org` (Internet). The subject line of the message should include the publication code to identify it as a Memoir. TeX source files, DVI files, and PostScript files can be transferred over the Internet by FTP to the Internet node `e-math.ams.org` (130.44.1.100).

Electronic graphics. Comprehensive instructions on preparing graphics are available at `www.ams.org/jourhtml/graphics.html`. A few of the major requirements are given here.

Submit files for graphics as EPS (Encapsulated PostScript) files. This includes graphics originated via a graphics application as well as scanned photographs or other computer-generated images. If this is not possible, TIFF files are acceptable as long as they can be opened in Adobe Photoshop or Illustrator. No matter what method was used to produce the graphic, it is necessary to provide a paper copy to the AMS.

Authors using graphics packages for the creation of electronic art should also avoid the use of any lines thinner than 0.5 points in width. Many graphics packages allow the user to specify a "hairline" for a very thin line. Hairlines often look acceptable when proofed on a typical laser printer. However, when produced on a high-resolution laser imagesetter, hairlines become nearly invisible and will be lost entirely in the final printing process.

Screens should be set to values between 15% and 85%. Screens which fall outside of this range are too light or too dark to print correctly. Variations of screens within a graphic should be no less than 10%.

Inquiries. Any inquiries concerning a paper that has been accepted for publication should be sent directly to the Electronic Prepress Department, American Mathematical Society, P. O. Box 6248, Providence, RI 02940-6248.

Editors

This journal is designed particularly for long research papers, normally at least 80 pages in length, and groups of cognate papers in pure and applied mathematics. Papers intended for publication in the *Memoirs* should be addressed to one of the following editors. In principle the Memoirs welcomes electronic submissions, and some of the editors, those whose names appear below with an asterisk (*), have indicated that they prefer them. However, editors reserve the right to request hard copies after papers have been submitted electronically. Authors are advised to make preliminary email inquiries to editors about whether they are likely to be able to handle submissions in a particular electronic form.

Algebra to KAREN E. SMITH, Department of Mathematics, University of Michigan, 525 University, Suite 2832, Ann Arbor, MI 48109-1109; email: kesmith@math.lsa.umich.edu

Algebraic geometry and commutative algebra to LAWRENCE EIN, Department of Mathematics, University of Illinois, 851 S. Morgan (M/C 249), Chicago, IL 60607-7045; email: ein@uic.edu

Algebraic topology and cohomology of groups to STEWART PRIDDY, Department of Mathematics, Northwestern University, 2033 Sheridan Road, Evanston, IL 60208-2730; email: priddy@math.nwu.edu

Combinatorics and Lie theory to SERGEY FOMIN, Department of Mathematics, University of Michigan, Ann Arbor, Michigan 48109-1109; email: fomin@math.lsa.umich.edu

Complex analysis and complex geometry to DUONG H. PHONG, Department of Mathematics, Columbia University, 2990 Broadway, New York, NY 10027-0029; email: phong@math.columbia.edu

***Differential geometry and global analysis** to LISA C. JEFFREY, Department of Mathematics, University of Toronto, 100 St. George St., Toronto, ON Canada M5S 3G3; email: jeffrey@math.toronto.edu

***Dynamical systems and ergodic theory** to ROBERT F. WILLIAMS, Department of Mathematics, University of Texas, Austin, Texas 78712-1082; email: bob@math.utexas.edu

Functional analysis and operator algebras to DAN VOICULESCU, Department of Mathematics, University of California, Berkeley, 970 Evans Hall, Floor 9, Berkeley, CA 94720-0001; email: dvv@math.berkeley.edu

Geometric topology, knot theory and hyperbolic geometry to ABIGAIL A. THOMPSON, Department of Mathematics, University of California, Davis, Davis, CA 95616-5224; email: thompson@math.ucdavis.edu

Harmonic analysis, representation theory, and Lie theory to ROBERT J. STANTON, Department of Mathematics, The Ohio State University, 231 West 18th Avenue, Columbus, OH 43210-1174; email: stanton@math.ohio-state.edu

***Logic** to THEODORE SLAMAN, Department of Mathematics, University of California, Berkeley, CA 94720-3840; email: slaman@math.berkeley.edu

Number theory to HAROLD G. DIAMOND, Department of Mathematics, University of Illinois, 1409 W. Green St., Urbana, IL 61801-2917; email: diamond@math.uiuc.edu

***Ordinary differential equations, partial differential equations, and applied mathematics** to PETER W. BATES, Department of Mathematics, Brigham Young University, 292 TMCB, Provo, UT 84602-1001; email: peter@math.byu.edu

***Probability and statistics** to KRZYSZTOF BURDZY, Department of Mathematics, University of Washington, Box 354350, Seattle, Washington 98195-4350; email: burdzy@math.washington.edu

***Real and harmonic analysis and geometric partial differential equations** to WILLIAM BECKNER, Department of Mathematics, University of Texas, Austin, TX 78712-1082; email: beckner@math.utexas.edu

All other communications to the editors should be addressed to the Managing Editor, WILLIAM BECKNER, Department of Mathematics, University of Texas, Austin, TX 78712-1082; email: beckner@math.utexas.edu.

Selected Titles in This Series

(Continued from the front of this publication)

706 **Jonathan Brundan, Richard Dipper, and Alexander Kleshchev,** Quantum Linear groups and representations of $GL_n(\mathbb{F}_q)$, 2001

705 **I. Moerdijk and J. J. C. Vermeulen,** Proper maps of toposes, 2000

704 **Jeff Hooper, Victor Snaith, and Min van Tran,** The second Chinburg conjecture for quaternion fields, 2000

703 **Erik Guentner, Nigel Higson, and Jody Trout,** Equivariant E-theory for C^*-algebras, 2000

702 **Ilijas Farah,** Analytic guotients: Theory of liftings for quotients over analytic ideals on the integers, 2000

701 **Paul Selick and Jie Wu,** On natural coalgebra decompositions of tensor algebras and loop suspensions, 2000

700 **Vicente Cortés,** A new construction of homogeneous quaternionic manifolds and related geometric structures, 2000

699 **Alexander Fel′shtyn,** Dynamical zeta functions, Nielsen theory and Reidemeister torsion, 2000

698 **Andrew R. Kustin,** Complexes associated to two vectors and a rectangular matrix, 2000

697 **Deguang Han and David R. Larson,** Frames, bases and group representations, 2000

696 **Donald J. Estep, Mats G. Larson, and Roy D. Williams,** Estimating the error of numerical solutions of systems of reaction-diffusion equations, 2000

695 **Vitaly Bergelson and Randall McCutcheon,** An ergodic IP polynomial Szemerédi theorem, 2000

694 **Alberto Bressan, Graziano Crasta, and Benedetto Piccoli,** Well-posedness of the Cauchy problem for $n \times n$ systems of conservation laws, 2000

693 **Doug Pickrell,** Invariant measures for unitary groups associated to Kac-Moody Lie algebras, 2000

692 **Mara D. Neusel,** Inverse invariant theory and Steenrod operations, 2000

691 **Bruce Hughes and Stratos Prassidis,** Control and relaxation over the circle, 2000

690 **Robert Rumely, Chi Fong Lau, and Robert Varley,** Existence of the sectional capacity, 2000

689 **M. A. Dickmann and F. Miraglia,** Special groups: Boolean-theoretic methods in the theory of quadratic forms, 2000

688 **Piotr Hajłasz and Pekka Koskela,** Sobolev met Poincaré, 2000

687 **Guy David and Stephen Semmes,** Uniform rectifiability and quasiminimizing sets of arbitrary codimension, 2000

686 **L. Gaunce Lewis, Jr.,** Splitting theorems for certain equivariant spectra, 2000

685 **Jean-Luc Joly, Guy Metivier, and Jeffrey Rauch,** Caustics for dissipative semilinear oscillations, 2000

684 **Harvey I. Blau, Bangteng Xu, Z. Arad, E. Fisman, V. Miloslavsky, and M. Muzychuk,** Homogeneous integral table algebras of degree three: A trilogy, 2000

683 **Serge Bouc,** Non-additive exact functors and tensor induction for Mackey functors, 2000

682 **Martin Majewski,** ational homotopical models and uniqueness, 2000

681 **David P. Blecher, Paul S. Muhly, and Vern I. Paulsen,** Categories of operator modules (Morita equivalence and projective modules, 2000

680 **Joachim Zacharias,** Continuous tensor products and Arveson's spectral C^*-algebras, 2000

For a complete list of titles in this series, visit the AMS Bookstore at **www.ams.org/bookstore/**.